1 MONTH OF
FREE
READING

at

www.ForgottenBooks.com

By purchasing this book you are eligible for one month membership to ForgottenBooks.com, giving you unlimited access to our entire collection of over 1,000,000 titles via our web site and mobile apps.

To claim your free month visit:

www.forgottenbooks.com/free46378

ISBN 978-1-5284-5435-3
PIBN 10046378

THE

NATURAL ARITHMETIC

BOOK II

BY

ISAAC O. WINSLOW, M.A.

PRINCIPAL OF THAYER STREET GRAMMAR SCHOOL
PROVIDENCE, R.I.

NEW YORK ·:· CINCINNATI ·:· CHICAGO
AMERICAN BOOK COMPANY

PURPOSES AND DISTINCTIVE FEATURES

THE purposes of this series of Arithmetics are : —

1. *To present the subjects in a spiral order.*

Instead of presenting the general subjects of addition, subtraction, multiplication, division, fractions, etc., as complete wholes in regular succession, each subject is divided into parts with reference to the difficulty of the principles involved. The easier principles of various subjects are treated together, while the more difficult principles are reserved until the child has gained the power to apprehend them easily.

2. *To make the work easy.*

In the belief that it is better to keep mathematical work a little behind the child's mental grasp than to advance it beyond that limit, the work designed for the different grades has been made somewhat easier than that usually found in text-books. The pupil is kept busy with a varied application of the principles that he has already mastered instead of being too rapidly crowded forward into greater difficulties.

3. *To give the subject variety and interest.*

The problems are based upon facts and principles gathered from the different branches of study, as history, geography, nature study, astronomy, and physics, as well as on the customary commercial transactions, thus correlating arithmetic with other studies and adding distinctly to its vividness and interest.

3

4. *To develop genuine mathematical thought.*

There is a large amount of mental work interspersed with the written work. Each new subject is first developed with numbers that are not too large for mental solution. After the principle has been well established, written practice with larger numbers is introduced. Originality is also promoted by exercises requiring the pupils to make problems for themselves from given data.

5. *To give prominence to the idea of magnitude.*

The psychological fact that all mathematical knowledge is a system of relations, or ratios, has been recognized throughout the series. For the purpose of suggesting to the pupils the application of number to magnitude, the simple geometric forms have been gradually introduced.

Book II takes up the development of the subject at the point where Book I ends, but in order to avoid too great dependence upon the earlier book the more important subjects in Book I are briefly reviewed.

The principles of common fractions, decimal fractions, and percentage are gradually developed in their simpler aspects, the more difficult parts being reserved for Book III.

The method with each subject is first to lead the child to a genuine grasp of the conceptions involved, withholding until later a formal statement of the processes. Considerable use is made of diagrams representing quantity, which are designed both as aids to correct thinking and as stimulants to the exercise of mathematical imagination.

INDEX

The numbers refer to pages.

5

Index

6

Index

BOOK II

Notation and Numeration

Quadrillions	Trillions	Billions	Millions	Thousands	Units

15,327,135,054,860,246

Read the following numbers:

1.	2.	3.
8024	1,000,000	200,000,000
30,105	5,008,024	560,036,015
508,096	40,020,500	9,000,400,200

4. Write twenty thousand, one hundred ten.

5. Write three hundred thousand, nine hundred.

6. Write five million, five hundred thousand.

7. Write fifty million, two hundred fifty thousand.

8. Write two hundred fifty million, eight thousand.

9. Write one hundred eight million, sixty thousand, five hundred.

10. Write four billion, twenty-six million, twelve thousand.

11. Write one billion, ten million, one hundred thousand.

9

Addition

Prove the addition by adding downwards as well as upwards.

Add:

1.	2.	3.	4.
2543	3281	9872	2751
6091	465	3654	3689
785	573	148	4522
5932	4907	7621	86
860	1223	95	9127
27	6580	872	375
390	25	6049	8422
6758	932	582	346
3299	5090	4233	9427
467	87	842	2146
1552	941	8522	892
879	3246	9650	978
1200	942	856	1240
5324	2508	8139	7136

Write in columns and add:

5. Two hundred forty-two thousand, five; ninety thousand, one hundred ninety; one hundred forty thousand, twenty; three million, twenty-five thousand, seven hundred sixty-three.

6. Twenty million, three hundred five thousand, one hundred; seventy million, seventy thousand, seven; eight hundred four thousand, one hundred; one million, two hundred twenty-two.

7. One million, one thousand, one hundred; fifty-five million, four hundred five thousand, twenty-eight; twenty thousand, three; two hundred five million, three hundred thousand, three hundred fifty-six.

Problems

1. In 1900 New York had a population of 3,437,202 and Chicago 1,698,575. How much greater was the population of New York than that of Chicago?

2. The gold mined in the United States in 1894 was valued at $39,500,000 and that mined in Africa in the same year at $40,271,000. How much more was procured in Africa than in the United States?

3. Mt. Everest, the highest mountain in the world, is 29,002 feet high. Pikes Peak in Colorado is 14,147 feet high. How much higher is Mt. Everest than Pikes Peak?

4. How much more is the sum of 7586 and 5420 than the sum of 6942 and 5315?

5. How much more is the sum of 42,521 and 12,362 than the difference between 59,532 and 15,361?

6. How many bushels of potatoes will 160 acres yield, if each acre yields 240 bushels?

7. How many pounds will 748 cubic feet of ice weigh, if one cubic foot weighs 58 pounds?

8. The area of Rhode Island is 1250 square miles. Its population in 1900 was 428,556. How many people were there per square mile?

9. There are 5280 feet in a mile. How many feet is it from New York to Albany, a distance of 143 miles?

10. There are 63,360 inches in a mile. How many steps, 30 inches long, must a man take in walking a mile?

11. There are 640 acres in a square mile. How many square miles are there in the state of Ohio, which contains 26,278,400 acres?

11

Drill Work

Subtract :

1. 256 from 9385.
2. 309 from 8500.
3. 573 from 7000.
4. 299 from 5600.
5. 188 from 3050.
6. 900 from 2100.
7. 788 from 3110.
8. 496 from 5685.

9. 2135 from 6582.
10. 3040 from 5101.
11. 1223 from 4021.
12. 5216 from 8758.
13. 2445 from 5444.
14. 1009 from 8000.
15. 2011 from 7010.
16. 3112 from 4111.

Multiply :

17. 325 by 324.
18. 687 by 156.
19. 703 by 225.
20. 623 by 344.
21. 450 by 256.
22. 648 by 320.
23. 670 by 470.
24. 503 by 890.

25. 5432 by 256.
26. 8015 by 742.
27. 7030 by 426.
28. 9350 by 580.
29. 4070 by 600.
30. 6000 by 580.
31. 8050 by 402.
32. 7000 by 900.

Divide :

33. 17,286 by 201.
34. 19,392 by 606.
35. 14,112 by 252.
36. 16,398 by 911.
37. 12,992 by 406.
38. 18,420 by 307.
39. 18,538 by 713.
40. 20,124 by 258.

41. 19,920 by 415.
42. 21,873 by 317.
43. 36,900 by 246.
44. 21,978 by 666.
45. 19,744 by 2468.
46. 36,630 by 1665.
47. 16,284 by 2714.
48. 29,616 by 7404.

Mental Problems

1. If two yards of cloth cost 40 cents, how much will one yard cost? How much will three yards cost?

2. If a boat sails 72 miles in six hours, how far will it sail in one hour? How far in five hours?

3. If a family eat six bushels of potatoes in two months, how many bushels will they eat in ten months?

4. How many oranges wil it take to give seven boys four oranges each, and have five oranges left?

5. How many oranges will it take for three boys and two girls, if each boy receives four oranges and each girl five oranges?

6. If a boy can distribute 60 papers in two hours, how many papers can three boys distribute in one hour?

7. If I buy apples at the rate of three for four cents, how much will twelve apples cost?

8. How much will fifteen apples cost?

9. If I exchange butter at twenty cents a pound for sugar at five cents a pound, how many pounds of sugar shall I receive for four pounds of butter?

10. A man bought seven oranges for six cents each and sold them for 54 cents. How much did he gain?

11. If a five acre lot is divided into ten equal parts, how much is one of the parts worth at $40 an acre?

12. If I buy a pound and a half of butter at 24 cents a pound, and half a dozen eggs at seventeen cents a dozen, and give a dollar bill in payment, how much change shall I receive?

13. If I buy two pounds of meat at eleven cents a pound, and a pound and a half of cheese at sixteen cents a pound, and pay with a two-dollar bill, how much change shall I receive?

Halves — Fourths — Eighths

This work should be wholly mental. Refer to the diagram, whenever necessary.

1. How many halves are there in a whole? In three wholes?

2. How many halves in two wholes and one half?

3. How many fourths in one and one fourth?

4. How many eighths in one half and one eighth?

5. How many eighths in three wholes and three eighths?

6. How many eighths in one whole, one half, and one fourth?

7. How much is three fourths and one half?

8. How much is three fourths and three eighths?

9. How much is one half less one eighth?

10. How much is seven eighths less one half?

A part of a whole is called a **fraction**.

When a fraction is written, the figure below the line indicates the number of parts into which the whole is regarded as divided and is called the **denominator**. The figure above the line indicates how many of these parts there are in the fraction and is called the **numerator**.

11. $\frac{1}{2} + \frac{1}{4}$ **14.** $\frac{1}{2} + \frac{1}{8}$ **17.** $\frac{1}{4} + \frac{1}{8}$ **20.** $\frac{7}{8} - \frac{1}{2}$

12. $\frac{1}{4} + \frac{3}{4}$ **15.** $\frac{1}{2} + \frac{3}{8}$ **18.** $\frac{3}{4} + \frac{1}{8}$ **21.** $\frac{3}{4} - \frac{5}{8}$

13. $\frac{3}{4} + \frac{3}{2}$ **16.** $\frac{7}{8} + \frac{1}{2}$ **19.** $\frac{7}{8} + \frac{1}{4}$ **22.** $\frac{3}{2} - \frac{7}{8}$

14

Fractions — Addition and Subtraction

1. Add $3\frac{1}{4}$, $2\frac{1}{2}$, and $5\frac{3}{4}$.

$3\frac{1}{4}$
$2\frac{1}{2}$ $\frac{1}{4}$ and $\frac{1}{2}$ are $1\frac{1}{4}$. $1\frac{1}{4}$ and $\frac{1}{4}$ are $1\frac{1}{2}$.
$5\frac{3}{4}$ Write the $\frac{1}{2}$ and add the 1 to the column of wholes.
$\overline{11\frac{1}{2}}$

2. From $8\frac{1}{4}$ subtract $3\frac{1}{2}$.

$8\frac{1}{4}$ $\frac{1}{2}$ cannot be taken from $\frac{1}{4}$.   We take one whole from the 8
$3\frac{1}{2}$ wholes, leaving 7 wholes.   This whole with the $\frac{1}{4}$ makes $\frac{5}{4}$.
$\overline{4\frac{3}{4}}$ $\frac{1}{2}$ from $\frac{5}{4}$ leaves $\frac{3}{4}$.   3 from 7 leaves 4.

Add:

3.	4.	5.	6.
$5\frac{1}{2}$	$15\frac{1}{8}$	$12\frac{1}{2}$	$22\frac{1}{8}$
$8\frac{1}{4}$	$6\frac{3}{8}$	$9\frac{3}{8}$	$36\frac{1}{4}$
$6\frac{3}{4}$	$8\frac{1}{4}$	$10\frac{3}{4}$	$42\frac{7}{8}$
$4\frac{1}{2}$	$7\frac{5}{8}$	$6\frac{5}{8}$	$50\frac{1}{2}$
$9\frac{3}{4}$	$10\frac{3}{4}$	$11\frac{1}{2}$	$63\frac{3}{4}$

Subtract:

7.	8.	9.	10.
$9\frac{1}{2}$	$10\frac{1}{4}$	$12\frac{7}{8}$	$14\frac{3}{8}$
$6\frac{1}{4}$	$5\frac{3}{4}$	$8\frac{3}{8}$	$9\frac{5}{8}$

11.	12.	13.	14.
$25\frac{1}{2}$	$36\frac{1}{4}$	$45\frac{3}{8}$	$64\frac{1}{4}$
$11\frac{1}{8}$	$21\frac{3}{4}$	$21\frac{3}{4}$	$20\frac{3}{8}$

15. How many gallons are $3\frac{1}{4}$ gallons, $5\frac{1}{2}$ gallons, and $2\frac{3}{8}$ gallons?

16. How many miles are $5\frac{1}{8}$ miles, $4\frac{1}{2}$ miles, and $2\frac{3}{4}$ miles?

Mental Problems

1. Five is $\frac{1}{4}$ of what number?

5 is $\frac{1}{4}$ of 4 times 5.

2. Six is $\frac{3}{4}$ of what number?

Since 6 is 3 fourths of the number, 1 fourth of the number is $\frac{1}{3}$ of 6 or 2. 4 fourths or the whole number is 4×2 or 8. Therefore, 6 is $\frac{3}{4}$ of 8.

3. Ten is $\frac{5}{8}$ of what number?

4. If $\frac{3}{4}$ of a yard of cloth costs six cents, how much will a whole yard cost?

5. If $\frac{3}{4}$ of a yard of cloth costs 9 cents, how much will half a yard cost?

6. If a quart and a half of milk cost 9 cents, how much will half a quart cost?

7. If 12 is $\frac{3}{8}$ of some number, what is $\frac{1}{8}$ of the number?

8. If 12 is $\frac{3}{8}$ of some number, what is the number?

9. If a boy gives away $\frac{3}{4}$ of his apples and has two left, how many did he have at first?

10. If a boy sells $\frac{7}{8}$ of his papers and has three left, how many did he have at first?

11. John is $\frac{3}{4}$ as old as his brother William. William is 16 years old. How old is John?

12. How many wholes are there in 20 fourths?

13. How many wholes in 24 eighths?

14. How many eighths in 5 wholes and 3 eighths?

15. How much will two barrels of flour cost at $4\frac{1}{2}$ a barrel?

16. If one half of an acre of land costs $15, how much will five acres cost?

17. If a farmer has $20\frac{1}{2}$ bushels of apples and sells $10\frac{3}{4}$ bushels, how many bushels will he have left?

16

Miscellaneous Problems

1. How many pints are there in a gallon?

2. How many pints in 1 gallon, 2 quarts, and 1 pint?

3. If a milkman should deliver 2 quarts of milk to each customer, how many customers would he supply in delivering 8½ gallons of milk?

4. How many quarts are there in a bushel?

5. How many quarts in 1 bushel and 3 pecks?

6. If I should sell ¼ of a bushel of berries at 5 cents a quart, how much should I receive?

7. If I should buy half a bushel of berries for 50 cents, and should sell them at 5 cents a quart, how much should I gain?

8. What part of a peck is 1 quart?

9. What part of a half bushel are 2 quarts?

10. If I should buy a peck of apples for 48 cents, and should sell them for 64 cents, how much should I make on each quart?

11. A man having 4 horses gives each horse 4 quarts of oats a day. How long will five bushels last?

12. If 3 quarts of oil cost 9 cents, how much is that a gallon?

13. If 2 quarts of beans cost 12 cents, how much is that a peck?

14. If 9 is $\frac{3}{8}$ of some number, what is $\frac{5}{8}$ of the number?

15. If 3 men can do a piece of work in 4 days, how many men will it take to do the work in 2 days?

16. If 8 men can harvest an acre of potatoes in 2 days, how many acres can they harvest in 10 days?

Fractions — Multiplication and Division

Let this work be entirely mental.

1. How much is four times one half? Six times one eighth? Five times one fourth? Eight times three fourths?

2. What is one fourth of twenty? Three halves of sixteen? Five eighths of forty?

3. How many times is one fourth contained in one half? One eighth in one half? One eighth in one fourth? One fourth in three fourths?

4. Divide three fourths by three. Nine eighths by three. Four halves by two.

5. How much is 6 times $4\frac{1}{2}$?

Multiply the whole number and the fractional part separately. 6 times 4 are 24. 6 times $\frac{1}{2}$ are $\frac{6}{2}$, or 3. 6 times $4\frac{1}{2}$ are 27.

6. Divide $6\frac{3}{4}$ by 3.

One third of 6 is 2 and $\frac{1}{3}$ of $\frac{3}{4}$ is $\frac{1}{4}$. 3 is contained $2\frac{1}{4}$ times in $6\frac{3}{4}$.

7. Divide $3\frac{1}{2}$ by $\frac{1}{4}$.

$\frac{1}{4}$ is contained in 3 wholes 12 times and in $\frac{1}{2}$ twice. $\frac{1}{4}$ is contained 14 times in $3\frac{1}{2}$.

8. $\frac{1}{2} \times 4$ **11.** $4 \times 1\frac{1}{4}$ **14.** $6\frac{3}{4} \div 3$ **17.** $2\frac{1}{2} \div \frac{1}{2}$

9. $\frac{3}{4} \times 8$ **12.** $8 \times 2\frac{3}{8}$ **15.** $28\frac{7}{8} \div 7$ **18.** $1\frac{1}{4} \div \frac{1}{8}$

10. $\frac{5}{8} \times 24$ **13.** $5 \times 5\frac{2}{5}$ **16.** $30\frac{3}{4} \div 3$ **19.** $4\frac{3}{4} \div \frac{1}{4}$

20. At $\frac{1}{2}$ of a cent each, how many apples can be bought for $6\frac{1}{2}$ cents?

21. How much will $\frac{3}{4}$ of a pound of tea cost at 40 cents a pound?

22. At the rate of a quarter of a dollar a pound, how many pounds of butter can be bought for $5\frac{1}{4}$ dollars?

23. If six boys have each $\frac{3}{4}$ of a dollar, how many dollars have they all?

18

1. Add $\frac{1}{2}$ and $\frac{3}{8}$.

2. Add $\frac{3}{4}$ and $\frac{1}{2}$.

3. Add $\frac{7}{8}$ and $\frac{3}{4}$.

4. Add $2\frac{1}{2}$ and $4\frac{3}{4}$.

5. Add $5\frac{1}{4}$ and $3\frac{5}{8}$.

6. Add $8\frac{3}{4}$ and $7\frac{3}{8}$.

7. $\frac{3}{4} + \frac{1}{2} + \frac{1}{8} = ?$

8. $\frac{3}{8} + \frac{1}{4} + \frac{1}{2} = ?$

9. $\frac{5}{8} + \frac{3}{4} + \frac{1}{8} = ?$

10. $1\frac{1}{4} + 2\frac{1}{2} + 2\frac{1}{4} = ?$

11. $3 + 4\frac{1}{4} + 5\frac{1}{2} = ?$

12. $6\frac{3}{4} + 2\frac{1}{2} + 3\frac{5}{8} = ?$

13. From $\frac{1}{2}$ take $\frac{1}{8}$.

14. From $\frac{3}{4}$ take $\frac{3}{8}$.

15. From 3 take $\frac{3}{4}$.

16. From $2\frac{1}{2}$ take $\frac{5}{8}$.

17. From $6\frac{3}{4}$ take $2\frac{3}{8}$.

18. From $2\frac{1}{4}$ take $\frac{5}{8}$.

19. $\frac{7}{8} - \frac{3}{8} = ?$

20. $\frac{5}{8} - \frac{1}{4} = ?$

21. $1\frac{1}{8} - \frac{1}{2} = ?$

22. $2\frac{1}{2} - 1\frac{1}{4} = ?$

23. $6\frac{3}{4} - 2\frac{3}{8} = ?$

24. $12\frac{1}{2} - 8\frac{3}{8} = ?$

25. Multiply $\frac{1}{4}$ by 3.

26. Multiply $\frac{1}{4}$ by 6.

27. Multiply $\frac{3}{8}$ by 5.

28. Multiply $3\frac{1}{4}$ by 4.

29. Multiply $5\frac{1}{8}$ by 6.

30. Multiply $4\frac{3}{4}$ by 3.

31. $\frac{1}{2} \times 3 = ?$

32. $\frac{3}{4} \times 5 = ?$

33. $\frac{7}{8} \times 6 = ?$

34. $2\frac{1}{2} \times 4 = ?$

35. $4\frac{1}{4} \times 8 = ?$

36. $5\frac{3}{8} \times 6 = ?$

37. Divide $\frac{3}{4}$ by 3.

38. Divide $\frac{1}{4}$ by $\frac{1}{8}$.

39. Divide $\frac{3}{4}$ by $\frac{1}{8}$.

40. Divide $2\frac{1}{2}$ by $\frac{1}{2}$.

41. Divide $1\frac{1}{4}$ by $\frac{1}{4}$.

42. Divide $1\frac{1}{2}$ by $\frac{3}{8}$.

43. $\frac{3}{8} \div 3 = ?$

44. $\frac{3}{2} \div \frac{3}{4} = ?$

45. $\frac{7}{8} \div \frac{1}{8} = ?$

46. $2\frac{1}{2} \div \frac{1}{4} = ?$

47. $3\frac{1}{2} \div \frac{1}{2} = ?$

48. $\frac{5}{8} \div 5 = ?$

Original Problems

Make problems from the following statements and solve them :

Let each pupil make a problem, then let some one state his problem for the class to solve. See who can make the best problem.

1. The sum of two numbers is 24 and one of the numbers is 14.

2. The sum of two numbers is 971.

3. The product of two numbers is 20 and one of the numbers is 5.

4. The product of two numbers is 250.

5. A basket contains 29 apples. 6 children are to receive 4 apples each.

6. A boy bought some oranges at the rate of 2 for 5 cents and sold them at 4 cents each.

7. A man took 8 doz. eggs to market and sold them at 12 cents a doz. He received his pay in meat at 8 cents a pound.

8. A man bought $\frac{7}{8}$ of an acre of land at the rate of $64 an acre.

9. $\frac{3}{4}$ of a certain number is 18.

10. A boy sold $\frac{5}{8}$ of his papers and had 6 left.

11. A farmer had $16\frac{3}{4}$ bushels of potatoes and sold $7\frac{7}{8}$ bushels.

12. A milkman had 20 customers and left each customer 3 pints of milk.

13. I bought 3 pecks of berries for $1.00 and sold them for 7 cents a quart.

14. 4 men can harvest an acre of corn in 8 days.

15. I can buy apples at the rate of 3 for 5 cents.

16. 8 boys have $\frac{3}{4}$ of a dollar each.

Fractions — Drill Work

Find the cost of:

1. $25\frac{1}{2}$ lb. of butter at $28\cent$ per lb.
2. $24\frac{3}{4}$ yd. of cloth at $2 per yard.
3. $52\frac{1}{2}$ bu. of wheat at $1.08 per bu.
4. 248 lb. of sugar at $5\frac{1}{8}\cent$ per lb.
5. 110 bbl. of flour at $4.75 per bbl.
6. 48 tons of hay at $16\frac{1}{4}$ per ton.
7. $76\frac{2}{8}$ yd. of silk at $3 per yard.
8. 128 sheep at $6\frac{3}{4}$ each.
9. 45 hats at $2.25 each.
10. 36 overcoats at $25.50 each.
11. $36\frac{5}{8}$ yd. of ribbon at $16\cent$ a yard.
12. 240 lb. of raisins at $8\frac{3}{4}\cent$ per lb.
13. $76\frac{1}{2}$ bu. of corn at $72\cent$ per bu.
14. 28 cows at $42\frac{1}{2}$ each.
15. 200 tons of hay at $12.50 per ton.
16. 32 hats at $1\frac{5}{8}$ each.
17. 300 lb. of butter at $24\frac{3}{4}\cent$ per lb.
18. 60 yd. of carpet at $87\frac{1}{2}\cent$ per yard.
19. 100 bottles of ink at $4\frac{3}{4}\cent$ each.
20. 64 quires of paper at $14\frac{1}{2}\cent$ a quire.
21. 4000 envelopes at $16\frac{1}{2}$ per thousand.
22. 160 yd. of cloth at $37\frac{3}{4}\cent$ per yard.
23. 500 lb. of sugar at $4\frac{3}{4}\cent$ per lb.
24. 48 pairs of shoes at $2.75 per pair.
25. 65 coats at $9.25 each.

Long Measure

1. How many inches are there in a yard?

2. What part of a foot is 3 inches?

3. How many feet in 2 yards and 2 feet?

4. How many inches in 5 feet and 4 inches?

5. How many inches in 1 yard, 2 feet, and 5 inches?

6. If a yard of ribbon should be cut into six-inch pieces, how many pieces would there be?

7. How many yards of ribbon must be bought to make badges for a class of 36 pupils, if each badge requires $\frac{1}{4}$ of a yard?

8. How many feet are there in a rod?

9. How many steps 3 feet long does a man take in walking a rod?

10. How many yards are there in a rod?

11. How many sticks 6 inches long could be made from a pole a rod long?

12. How many feet long is a fence which is 4 rods long?

Illustrate the following problems by diagrams.

13. If 2 men 20 miles apart walk toward each other, each walking 5 miles an hour, how long will it be before they will meet?

14. If 2 men 27 miles apart walk toward each other, one walking 5 miles an hour and the other 4 miles an hour, how long will it be before they will meet?

15. If 2 men start at the same point and travel in opposite directions, one 4 miles an hour and the other 3 miles an' hour, how far apart will they be in 5 hours?

Thirds — Sixths

1. One third is how many sixths?
2. One half is how many sixths?
3. How much is one third and one sixth?
4. How much is two thirds and one sixth?
5. What is the sum of one half and one sixth?
6. What is the sum of one half and one third?
7. What is the sum of two thirds and one half?
8. Find the sum of one half, one third, and one sixth.
9. Five sixths less one third equals what?

10. $\frac{1}{2}+\frac{1}{3}$	13. $\frac{3}{2}-\frac{5}{6}$	16. $\frac{3}{4}+\frac{5}{8}$	19. $\frac{11}{8}-\frac{3}{4}$
11. $\frac{2}{3}+\frac{5}{6}$	14. $\frac{3}{2}-\frac{2}{3}$	17. $\frac{7}{8}-\frac{1}{4}$	20. $\frac{2}{3}-\frac{1}{2}$
12. $\frac{2}{3}-\frac{1}{6}$	15. $\frac{5}{2}+\frac{1}{6}$	18. $\frac{7}{8}+\frac{7}{8}$	21. $\frac{7}{8}-\frac{1}{2}$

22. $4\frac{1}{2}+3\frac{1}{6}$	25. $7\frac{1}{2}+4\frac{3}{4}$	28. $6\frac{5}{6}-2\frac{1}{3}$	31. $12\frac{7}{8}+6\frac{1}{4}$
23. $5\frac{2}{3}+4\frac{1}{2}$	26. $5\frac{1}{2}+8\frac{1}{3}$	29. $8\frac{1}{2}+4\frac{2}{3}$	32. $11\frac{5}{6}-4\frac{1}{2}$
24. $6\frac{1}{3}+2\frac{1}{6}$	27. $8\frac{2}{3}-4\frac{1}{2}$	30. $10\frac{7}{8}-5\frac{3}{4}$	33. $10\frac{2}{3}+8\frac{2}{3}$

	34.	**35.**	**36.**	**37.**
Add :	$5\frac{3}{4}$	$7\frac{1}{3}$	$15\frac{1}{2}$	$67\frac{7}{8}$
	$6\frac{2}{3}$	$4\frac{5}{6}$	$24\frac{2}{3}$	$43\frac{3}{4}$
	$3\frac{1}{2}$	$8\frac{2}{3}$	$18\frac{1}{6}$	$98\frac{1}{2}$
	$9\frac{1}{4}$	$5\frac{1}{6}$	$32\frac{1}{2}$	$57\frac{3}{8}$

Subtract :

	38.	**39.**	**40.**	**41.**
	$8\frac{1}{2}$	$12\frac{5}{6}$	$10\frac{1}{2}$	$11\frac{3}{8}$
	$4\frac{1}{6}$	$9\frac{1}{2}$	$5\frac{5}{6}$	$10\frac{3}{4}$

Mental Problems

If necessary, in solving these problems, draw diagrams similar to those on pages 14 and 23.

1. If a boy sells $\frac{2}{3}$ and $\frac{1}{6}$ of his marbles, what part has he left?

2. If a boy sells $\frac{2}{3}$ and $\frac{1}{6}$ of his marbles and has 4 left, how many had he at first?

$4 = \frac{1}{6}$ of what number?

3. If two thirds of a yard of cloth costs 10 cents, how much does a whole yard cost?

How much does $\frac{1}{3}$ of a yard cost? How much does $\frac{3}{3}$ of a yard cost?

4. If $\frac{5}{6}$ of a yard costs 15 cents, how much does half a yard cost?

First find the cost of a yard.

5. How many inches are there in $3\frac{1}{2}$ feet? $5\frac{1}{3}$ feet? $4\frac{1}{6}$ feet?

6. If I have $\$10\frac{1}{4}$ and spend $\$5\frac{1}{2}$, how many dollars shall I have left?

7. A man sold a watch for $\$10$, which was $\frac{5}{6}$ of what it cost him. What was the cost?

8. A man sold a watch for $\$12$, which was $\frac{2}{3}$ of what it cost him. How much did he lose?

First find the cost.

9. If $\frac{1}{3}$ of the time of the forenoon school session is devoted to reading, $\frac{1}{4}$ to arithmetic, $\frac{1}{6}$ to writing, and the remainder to spelling, what part of the time is devoted to spelling?

10. If $1\frac{1}{3}$ acres of land cost $\$40$, how much is that an acre?

11. If $2\frac{1}{2}$ tons of coal cost $\$15$, how much does half a ton cost?

24

1. Multiply 16 by $4\frac{3}{4}$.

$$\begin{array}{r} 16 \\ 4\frac{3}{4} \\ \hline 12 \\ 64 \\ \hline 76 \end{array}$$

$\frac{3}{4}$ times 16 is the same as $\frac{3}{4}$ of 16. $\frac{1}{4}$ of $16 = 4$; $\frac{3}{4}$ of $16 = 12$. 4 times $16 = 64$. $4\frac{3}{4}$ times $16 = 76$.

2. Divide $8\frac{1}{2}$ by $4\frac{1}{4}$.

$$4\frac{1}{4})\overline{8\frac{1}{2}}$$
$$2$$

$4\frac{1}{4}$ is contained in $8\frac{1}{2}$ twice, because twice $4\frac{1}{4}$ is $8\frac{1}{2}$.

Multiply :

3.	4.	5.	6.	7.	8.	9.
$12\frac{1}{2}$	$16\frac{3}{4}$	$25\frac{3}{8}$	$32\frac{5}{6}$	36	48	12
8	4	6	3	$5\frac{1}{2}$	$7\frac{1}{6}$	$8\frac{2}{3}$

Divide :

10.	11.	12.	13.	14.	15.	16.
$3)9\frac{3}{4}$	$5)15\frac{5}{6}$	$7)49\frac{7}{8}$	$6)72\frac{6}{8}$	$1\frac{1}{2})3$	$3\frac{1}{3})10$	$6\frac{1}{4})12\frac{1}{2}$

17. How much will $7\frac{1}{2}$ yards of cloth cost at 6 cents a yard?

18. How many feet are there in 6 rods?

19. How many feet are there in 4 rods, $3\frac{1}{3}$ yards?

20. Into how many pieces $2\frac{1}{2}$ feet long can a board $12\frac{1}{2}$ feet long be divided?

21. How many feet are there in $\frac{1}{4}$ of a rod?

22. How many inches are there in $\frac{1}{6}$ of a yard?

23. How many pieces of string $\frac{2}{3}$ of a yard long can be cut from a piece 4 yards long?

24. If $\frac{1}{2}$ of a barrel of flour is divided equally among 3 persons, what part of a barrel does each receive?

25. If 2 bushels of potatoes are divided equally among 6 persons, what part of a bushel does each receive?

26. Six feet make a fathom. What part of a fathom is a foot and a half?

Compound Quantities

1. Add 3 gal. 2 qt. to 6 gal. 3 qt.

3 gal.	2 qt.
6 gal.	3 qt.
9 gal.	5 qt.
10 gal.	1 qt.

First add the columns of quantities which are alike and afterwards change the sums, as far as possible, to larger measures. 5 quarts make 1 gallon and 1 quart over. Add the 1 gallon to the 9 gallons.

2. From 6 yd. 1 ft. 8 in. subtract 2 yd. 2 ft. 5 in.

6 yd.	1 ft.	8 in.
2 yd.	2 ft.	5 in.
3 yd.	2 ft.	3 in.

When necessary, change one of the larger measures to smaller measures. 2 feet cannot be taken from 1 foot. Change 1 yard to 3 feet, which with the 1 foot make 4 feet. 2 feet from 4 feet leave 2 feet. 2 yards from 5 yards leave 3 yards.

See the tables on page 248.

Add:

3.

4 gal.	1 qt.	1 pt.
5 gal.	2 qt.	1 pt.

4.

7 gal.	3 qt.	
4 gal.	2 qt.	1 pt.

5.

8 bu.	2 pk.	4 qt.
5 bu.	3 pk.	2 qt.
2 bu.	1 pk.	5 qt.

6.

9 bu.	3 pk.	7 qt.
10 bu.	2 pk.	5 qt.
7 bu.		2 qt.

Subtract:

7.

12 bu.	3 pk.	7 qt.
6 bu.	2 pk.	4 qt.

8.

5 yd.	2 ft.	10 in.
4 yd.	1 ft.	8 in.

9.

24 bu.	2 pk.	6 qt.
10 bu.	3 pk.	3 qt.

10.

4 yd.	2 ft.	6 in.
2 yd.	2 ft.	7 in.

Measurements

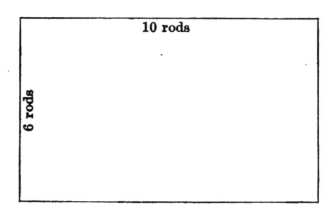

1. How many posts, 1 rod apart, will it take to go around a lot which is 10 rods long and 6 rods wide?

2. Find the distance in feet around the lot.

3. How many feet of wire will be required to make the fence, if it is built 3 wires high?

4. How many pounds of wire will be required, if 16 feet of wire weigh a pound?

5. What will be the cost of the posts and wire, if the posts cost 25 cents each and the wire $3\frac{1}{2}$ cents a pound?

6. If $5\frac{1}{2}$ paces make a rod, how many steps would a man take in walking around the lot?

7. How many times would it be necessary to walk around the lot in order to walk a mile?

8. Find how many yards there are in a mile.

9. Find how many feet there are in a mile.

10. How many minutes will it take to walk a mile, at the rate of 16 rods a minute?

11. If the distance around the wheel of a bicycle is 7 feet, how many times will the wheel turn around in going a mile?

12. What is the distance around a field which is 60 rods long and $22\frac{1}{2}$ rods wide?

Compound Quantities

Change:

1. 25 gal. to pints.
2. 5 bu. 3 pk. to pecks.
3. 6 ft. 4 in. to inches.
4. 18 gal. 3 qt. to pints.
5. 17 bu. 2 pk. to quarts.
6. 28 yd. 1 ft. to feet.
7. 456 pt. to gallons.
8. 640 qt. to bushels.
9. 756 in. to yards.
10. 323 qt. to gallons and quarts.
11. 175 pk. to bushels and pecks.
12. 250 ft. to yards and feet.
13. 7 gal. 3 qt. to pints.
14. 20 bu. 1 pk. to quarts.
15. 12 yd. 2 ft. 5 in. to inches.
16. 175 pt. to quarts and pints.
17. 269 qt. to pecks and quarts.
18. 100 in. to feet and inches.
19. 16 gal. 3 qt. 1 pt. to pints.
20. 5 bu. 3 pk. 7 qt. to quarts.
21. 750 in. to yards, feet, and inches.
22. 151 pt. to gallons, quarts, and pints.
23. 30 bu. 2 pk. 1 qt. to quarts.
24. 17 yd. 2 ft. 8 in. to inches.
25. 640 in. to yards, feet, and inches.

Fifths — Tenths

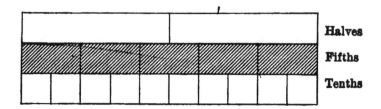

This work should be entirely mental.

1. One half is how many tenths?
2. One fifth is how many tenths?
3. How much is two fifths and one tenth?
4. How much is three fifths and one tenth?
5. How much is nine tenths and two fifths?
6. One half and one tenth is how much?
7. What is the sum of one half, one fifth, and one tenth?

8. $\frac{1}{2} + \frac{1}{5} + \frac{1}{10}$ 11. $3\frac{2}{5} + 4\frac{1}{10} + 2\frac{1}{2}$ 14. $\frac{1}{2} + \frac{2}{3} + \frac{1}{6}$

9. $\frac{3}{5} + \frac{2}{5} + \frac{1}{10}$ 12. $5\frac{1}{5} + 2\frac{7}{10} + 6$ 15. $\frac{1}{3} + \frac{5}{6} + \frac{1}{2}$

10. $\frac{9}{10} + \frac{1}{2} + \frac{4}{5}$ 13. $7\frac{9}{10} + 12 + 4\frac{1}{2}$ 16. $\frac{5}{6} + \frac{3}{2} + \frac{1}{3}$

17. $\frac{9}{10} - \frac{3}{5}$ 20. $4\frac{4}{5} - 2\frac{3}{10}$ 23. $5\frac{3}{4} - 2\frac{3}{8}$

18. $\frac{7}{10} - \frac{1}{2}$ 21. $6\frac{1}{2} - 3\frac{2}{5}$ 24. $9\frac{7}{8} - 4\frac{1}{4}$

19. $\frac{4}{5} - \frac{3}{10}$ 22. $8\frac{1}{10} - 5\frac{1}{5}$ 25. $7\frac{1}{8} - 3\frac{1}{4}$

In Example 30 say "one half of *two* thirds is *one* third."
In Example 38 say "one fourth is contained in one half twice."

Multiply:

26. $\frac{3}{5}$ by 5   28. 12 by $\frac{1}{2}$ 30. $\frac{2}{3}$ by $\frac{1}{2}$ 32. 10 by $3\frac{1}{5}$

27. $\frac{7}{10}$ by 2   29. 20 by $\frac{4}{5}$ 31. $\frac{4}{5}$ by $\frac{1}{4}$ 33. 15 by $2\frac{1}{3}$

Divide:

34. $\frac{4}{5}$ by 2   36. 4 by $\frac{1}{2}$ 38. $\frac{1}{2}$ by $\frac{1}{4}$.   40. $2\frac{1}{4}$ by $\frac{1}{4}$

35. $\frac{9}{10}$ by 3   37. 3 by $\frac{1}{5}$ 39. $\frac{2}{5}$ by $\frac{1}{10}$ 41. $5\frac{1}{2}$ by $\frac{1}{2}$

Mental Problems

1. How much is $\frac{2}{5}$ of 10 ? $\frac{1}{4}$ of 20 ? $\frac{2}{5}$ of 40 ?
2. How much is $\frac{3}{10}$ of 20 ? $\frac{7}{10}$ of 30 ? $\frac{9}{10}$ of 40 ?
3. What is $\frac{2}{8}$ of 24 ? $\frac{5}{8}$ of 32 ? $\frac{7}{8}$ of 48 ?
4. What is the difference between $\frac{2}{3}$ and $\frac{1}{2}$?
5. What is the difference between $\frac{2}{5}$ of 20 and $\frac{1}{4}$ of 20 ?
6. What is the sum of $\frac{1}{4}$ of 32 and $\frac{1}{8}$ of 32 ?
7. 9 is $\frac{3}{10}$ of what number ?
8. 15 is $\frac{5}{8}$ of what number ?
9. If 12 is $\frac{3}{4}$ of some number, what is $\frac{1}{8}$ of the same number ?
10. If 18 is $\frac{1}{2}$ of some number, what is $\frac{1}{3}$ of the same number ?
11. If $\frac{2}{5}$ of a bushel of wheat costs 36 cents, how much will a bushel cost ?
12. If $\frac{1}{4}$ of a bushel of wheat costs 20 cents, how much will $\frac{1}{8}$ of a bushel cost ?
13. How many times is $\frac{1}{10}$ of a bushel contained in $\frac{4}{5}$ of a bushel ?
14. If $\frac{3}{4}$ of a barrel of apples is divided equally among a number of persons so as to give each person $\frac{1}{8}$ of a barrel, how many persons are there ?
15. A boy having 28 apples kept 3 himself, and divided the remainder among 5 other boys. How many did each receive ?
16. Having 47 oranges, I sold 10 to one person, and 5 to another, and divided the remainder equally among 8 children. How many did each child receive ?
17. If I divide 15 sheets of music among an equal number of boys and girls, giving each boy 2 sheets and each girl 3 sheets, how many boys and how many girls are there ?

30

Fractions — Drill Work

1. Add $\frac{1}{2}$ and $\frac{2}{3}$.

2. Add $\frac{1}{3}$ and $\frac{5}{6}$.

3. Add $\frac{1}{6}$ and $\frac{1}{2}$.

4. Add $\frac{1}{2}$ and $\frac{2}{5}$.

5. Add $\frac{7}{8}$ and $\frac{3}{4}$.

6. Add $\frac{1}{2}$ and $\frac{9}{10}$.

7. $\frac{1}{2} + \frac{1}{3} + \frac{1}{6} = ?$

8. $\frac{3}{2} + \frac{1}{4} + \frac{3}{8} = ?$

9. $\frac{3}{5} + \frac{1}{2} + \frac{1}{10} = ?$

10. $3\frac{1}{3} + 4 + 5\frac{1}{6} = ?$

11. $2\frac{2}{5} + 1\frac{1}{2} + \frac{3}{10} = ?$

12. $4\frac{5}{6} + \frac{1}{6} + 2\frac{2}{3} = ?$

13. Take $\frac{1}{3}$ from $\frac{5}{6}$.

14. Take $\frac{1}{6}$ from $\frac{1}{2}$.

15. Take $\frac{1}{2}$ from $\frac{2}{3}$.

16. Take $\frac{3}{4}$ from $\frac{7}{8}$.

17. Take $\frac{1}{2}$ from $\frac{7}{10}$.

18. Take $\frac{4}{5}$ from $\frac{9}{10}$.

19. $\frac{1}{3} - \frac{1}{6} = ?$

20. $\frac{2}{5} - \frac{3}{10} = ?$

21. $\frac{5}{8} - \frac{1}{4} = ?$

22. $4\frac{1}{3} - 2\frac{1}{6} = ?$

23. $1\frac{5}{6} - \frac{1}{3} = ?$

24. $1\frac{1}{3} - \frac{5}{6} = ?$

25. Multiply $\frac{1}{3}$ by 6.

26. Multiply $\frac{5}{6}$ by 4.

27. Multiply $\frac{3}{5}$ by 5.

28. Multiply $\frac{3}{4}$ by 6.

29. Multiply $\frac{2}{3}$ by 9.

30. Multiply $\frac{7}{10}$ by 4.

31. $\frac{2}{5} \times 8 = ?$

32. $\frac{3}{2} \times 6 = ?$

33. $\frac{2}{3} \times 9 = ?$

34. $2\frac{1}{5} \times 5 = ?$

35. $4\frac{3}{10} \times 7 = ?$

36. $6\frac{5}{6} \times 3 = ?$

37. Divide $\frac{2}{3}$ by $\frac{1}{3}$.

38. Divide $\frac{3}{4}$ by $\frac{1}{8}$.

39. Divide $\frac{5}{6}$ by $\frac{1}{6}$.

40. Divide $\frac{4}{5}$ by $\frac{2}{5}$.

41. Divide $\frac{2}{3}$ by $\frac{1}{6}$.

42. Divide $\frac{9}{10}$ by $\frac{3}{10}$.

43. $\frac{7}{10} \div 7 = ?$

44. $\frac{4}{5} \div \frac{2}{5} = ?$

45. $\frac{2}{3} \div \frac{1}{6} = ?$

46. $3\frac{3}{5} \div 3 = ?$

47. $3\frac{3}{5} \div 1\frac{1}{5} = ?$

48. $6\frac{9}{10} \div 2\frac{3}{10} = ?$

Compound Quantities

1. Find the sum of 7 gal. 1 qt. 1 pt., 4 gal. 3 qt., and 2 qt. 1 pt.

2. Find the sum of 7 bu. 5 pk. 6 qt., 7 pk. 5 qt., and 21 bu. 3 pk.

3. Find the sum of 9 ft. 5 in., 17 yd. 9 in., and 35 yd. 2 ft.

4. Find the difference between 8 gal. 2 qt. 1 pt. and 5 gal. 3 qt. 1 pt.

5. Find the difference between 13 bu. 1 pk. 7 qt. and 9 bu. 3 pk. 4 qt.

6. Find the difference between 46 yd. 2 ft. 10 in. and 35 yd. 1 ft. 11 in.

7. A can contains 4 gal. 3 qt. of milk, a second can contains 2 gal. 2 qt. 1 pt., and a third 3 gal. 1 qt. 1 pt. How much milk is there in all?

8. From a barrel containing 29 gal. 2 qt. of oil 12 gal. 3 qt. 1 pt. leaked out. How much remained?

9. In a market there are 12 bu. 3 pk. of potatoes, 8 bu. 2 pk. 5 qt. of apples, 7 bu. 1 pk. 6 qt. of beets, and 3 pk. 4 qt. of turnips. How many bushels of vegetables are there in all?

10. Three strings measure separately 5 yd. 2 ft. 7 in., 7 yd. $2\frac{1}{2}$ ft. $5\frac{1}{4}$ in., and 10 yd. $1\frac{1}{2}$ feet, $6\frac{3}{4}$ in. What is the length of the three strings together?

11. A merchant bought three pieces of cloth containing 42 yd. 2 ft. 8 in., $22\frac{1}{2}$ yd. $1\frac{2}{3}$ ft., $15\frac{2}{3}$ yd. $2\frac{1}{4}$ ft. 4 in. How many yards did he buy in all?

12. From a piece of cloth containing 36 yd. 2 ft. a piece containing 15 yd. 1 ft. 6 in. has been sold. How much remains?

Original Problems

Make problems from these statements and solve them:

1. $2\frac{1}{2}$ yards of ribbon are to be cut into 6-inch pieces.

2. A certain field is 6 rods long. I can step 3 feet at each step.

3. Two men 12 miles apart walk toward each other.

4. A boy lost $\frac{1}{8}$ and $\frac{1}{6}$ of his marbles.

5. $\frac{5}{6}$ of an acre of land costs $50.

6. A man sold a watch for $\frac{3}{4}$ of what it cost him.

7. I sold $\frac{1}{2}$ of a barrel of apples to one man, and $\frac{2}{3}$ of a barrel to another.

8. A man can do $\frac{2}{3}$ of a piece of work in a day.

9. $\frac{1}{3}$ of an apple is divided equally between two children.

10. In each of 3 boxes there are 2 bu. 3 pk. 5 qt. of beans.

11. 4 bushels of apples are divided equally among 12 persons.

12. A merchant had a piece of cloth containing 42 yd. $2\frac{1}{2}$ ft. He sold 21 yd. $1\frac{1}{2}$ ft.

13. A can contained 6 gal. 2 qt. of milk.

14. It takes 8 flagstones, each 4 ft. 8 in. long, to reach across a street.

15. I can walk at the rate of 20 rods a minute.

16. A lot of land is 12 rods long and 8 rods wide.

17. $\frac{3}{5}$ of a bushel of corn costs 36 cents.

18. A boy had 17 apples and had 4 companions.

19. I had $\frac{7}{8}$ of a bushel of apples, and sold $\frac{3}{4}$ of a bushel.

Mental Problems

1. 7 is what part of 10?

2. 5 is what part of 15?

3. What part of $\frac{1}{2}$ is $\frac{1}{4}$?

4. What part of a gallon are 3 pints?

5. What part of a yard are 6 inches?

6. What part of a ton are 500 pounds?

7. 20 is $\frac{4}{5}$ of what number?

8. 60 is $\frac{3}{4}$ of what number?

9. At $5.00 a ton, what part of a ton of coal can be bought for $2.00?

10. If berries cost 5 cents a quart, what part of a peck can be bought for 25 cents?

11. 2 men hire a pasture for $20. One uses $\frac{7}{10}$ of it and the other $\frac{3}{10}$. How much should each man pay?

12. If a horse travels 4 miles in $\frac{2}{3}$ of an hour, how far will it travel in 2 hours?

13. If a boat sails $6\frac{3}{4}$ miles in an hour, how many miles will it sail in $\frac{1}{3}$ of an hour?

14. If I sell a coat for $8.00, losing $\frac{1}{5}$ of what it cost, what was the cost?

$\frac{5}{5} - \frac{1}{5} = \frac{4}{5}$. $8 is $\frac{4}{5}$ of what number?

15. If I sell a coat for $\frac{1}{6}$ more than the cost, and gain $3.00, what was the cost?

$3 is $\frac{1}{6}$ of what number?

16. If 25 cents of every dollar received by a merchant is profit, what part of what he receives is profit?

17. If 2 men start at the same point and travel in opposite directions, one traveling $4\frac{1}{2}$ miles an hour and the other $3\frac{1}{2}$ miles an hour, how far apart will they be in 5 hours?

34

Problems from Geography

Estimate the results approximately before solving the problems accurately.

1. From New York to San Francisco it is about 3270 miles. If a train should move continuously at the rate of 30 miles an hour, how many hours would it take to go the whole distance?

2. How many hours in all would it take, if 30 stops of 10 minutes each, should be made on the way?

3. How many days would it take a bicycle rider to go across the country, if he should ride 8 miles an hour upon the average, and 10 hours a day?

4. How many days would it take a vessel, sailing 12 miles an hour, to sail from San Francisco to the Hawaiian Islands, a distance of about 2000 miles?

5. The Erie Canal, built in 1825 from Albany to Buffalo, is 363 miles long. If a canal boat is moved at the rate of 3 miles an hour, how many hours will it take for a boat to make the trip?

6. If the boat moved 8 hours a day, at the rate of 3 miles an hour, how many days would it take?

7. Before the canal was built it cost $10 a barrel to transport flour from Albany to Buffalo. After the canal was built it cost only 30 cents a barrel. How much expense in freight was saved upon 100 barrels of flour by building the canal?

8. The distance between Boston and Denver is about 1400 miles. How many days would it take to walk that distance at the rate of 4 miles an hour, walking 8 hours each day?

9. Mt. Washington is 6286 feet high. Mt. Blanc is 15,744 feet high. How much higher is Mt. Blanc than Mt. Washington?

Avoirdupois Weight

1. How many ounces are there in a pound?

2. How many pounds in a ton?

3. A bushel of potatoes weighs 60 pounds. How many pounds are there in 2 bushels and 1 peck of potatoes?

4. A barrel of potatoes contains 165 pounds. How many bushels in a barrel of potatoes?

5. If there are 50 pounds in a bushel of meal, and a bag contains 2 bushels, what will be the weight of a load consisting of 45 bags?

6. How many horses would it take to draw such a load, if one horse cannot draw more than a ton?

7. If a basket of coal weighs 90 pounds, how many baskets would there be in a ton?

8. If baskets of coal average 80 pounds each, and the price is 25 cents a basket, how much is that a ton?

9. If coal is sold at $5.00 a ton, and a basket of coal weighs 80 pounds, what would be the price per basket at the same rate?

10. If 5 pounds of tea are divided into half-ounce sample packages, how many packages will there be?

11. If I buy nutmegs at 40 cents a pound and sell them at 4 cents an ounce, how much shall I make per pound?

12. A barrel of flour contains 196 pounds. Find how many loaves of bread can be made from a barrel of flour, if each loaf contains 14 ounces.

13. What part of a pound is 4 ounces? 12 ounces? 14 ounces?

14. What part of a ton is 200 pounds? 500 pounds? 1500 pounds?

15. What part of a ton is 250 pounds? 750 pounds? 1250 pounds?

			Thirds

(bar divided into thirds over ninths)

					Sixths

(bar divided into sixths over twelfths)

1. One third equals how many ninths? Two thirds?

2. One sixth equals how many twelfths?

3. Four twelfths equals how many sixths?

4. How much is one third and one ninth? Two thirds less one ninth? Two thirds and two ninths? Eight ninths less one third? Five ninths and two thirds? Two thirds less five ninths?

5. How much is one third and one twelfth? One sixth less one twelfth? One fourth and one twelfth? One third and one fourth? One fourth and one sixth?

6. $\frac{8}{9} + \frac{2}{3}$ **9.** $\frac{11}{9} - \frac{2}{3}$ **12.** $\frac{5}{4} + \frac{1}{6}$ **15.** $\frac{11}{12} + \frac{3}{4}$

7. $\frac{7}{9} - \frac{1}{3}$ **10.** $\frac{10}{9} + \frac{5}{3}$ **13.** $\frac{5}{6} - \frac{1}{4}$ **16.** $\frac{5}{6} + \frac{5}{4}$

8. $\frac{2}{3} + \frac{5}{9}$ **11.** $\frac{13}{9} - \frac{2}{3}$ **14.** $\frac{1}{4} - \frac{1}{6}$ **17.** $\frac{13}{12} - \frac{5}{6}$

Add :

18.	19.	20.	21.	22.
$3\frac{1}{2}$	$7\frac{3}{4}$	$10\frac{1}{3}$	$13\frac{1}{9}$	$22\frac{1}{12}$
4	$8\frac{1}{2}$	12	$16\frac{2}{3}$	$16\frac{1}{4}$
$7\frac{5}{8}$	$6\frac{1}{8}$	$9\frac{5}{6}$	$14\frac{5}{9}$	$30\frac{1}{6}$

Subtract :

23.	24.	25.	26.	27.
$10\frac{8}{9}$	$12\frac{4}{9}$	$25\frac{11}{12}$	$64\frac{3}{4}$	$45\frac{1}{4}$
$5\frac{2}{3}$	$8\frac{2}{3}$	$15\frac{3}{4}$	$22\frac{1}{3}$	$16\frac{1}{8}$

Multiplication and Division

Multiply:

1. $2\frac{1}{8}$ by 5   3. 6 by $5\frac{2}{3}$ 5. $24\frac{5}{6}$ by 8   7. $84\frac{7}{12}$ by 7

2. $5\frac{5}{8}$ by 4   4. 18 by $4\frac{5}{9}$ 6. $33\frac{3}{4}$ by 9   8. $65\frac{1}{2}$ by 15

Divide:

9. $20\frac{5}{6}$ by 5   11. 10 by $2\frac{1}{2}$ 13. $15\frac{5}{12}$ by 5   15. $4\frac{1}{8}$ by 3

10. $36\frac{2}{3}$ by 2   12. $4\frac{1}{2}$ by $1\frac{1}{2}$ 14. 10 by $3\frac{1}{8}$ 16. $6\frac{1}{4}$ by 5

17. How many are 9 times $6\frac{5}{7}$?

18. How many are 7 times $3\frac{5}{12}$?

19. How much will $7\frac{3}{8}$ pounds of sugar cost at 5 cents a pound ?

20. How much will $6\frac{9}{10}$ pounds of meat cost at 12 cents a pound ?

21. What is the value of $\frac{7}{10}$ of a ton of coal at \$8 a ton ?

22. What will be the cost of $9\frac{3}{8}$ yards of braid at 10 cents a yard ?

23. How many inches are there in $5\frac{11}{12}$ feet ?

24. 10 is $\frac{5}{6}$ of a number. What is $3\frac{3}{4}$ times the number ?

25. If 9 is $\frac{3}{4}$ of a number, what is $7\frac{1}{12}$ times the number ?

26. If $\frac{1}{6}$ of a yard of braid costs 2 cents, what will be the cost of $6\frac{5}{9}$ yards ?

27. If $\frac{2}{3}$ of a pound of rice costs 6 cents, how much will $10\frac{5}{6}$ pounds cost ?

28. If $\frac{3}{8}$ of a gallon of sirup costs 30 cents, how much will $5\frac{3}{4}$ gallons cost ?

29. 4 barrels of apples cost \$12. How much will $2\frac{1}{2}$ barrels cost ?

38

1. What part of 3 blocks are 2 blocks? What part of 3 fifths is 2 fifths?

1 is $\frac{1}{3}$ of 3; 2 is $\frac{2}{3}$ of 3.

2. What part of $\frac{3}{5}$ is $\frac{2}{5}$?

3. What part of $\frac{8}{9}$ is $\frac{5}{9}$?

4. If $\frac{2}{3}$ of a yard of ribbon costs 4 cents, how much will $\frac{1}{2}$ of a yard cost?

First find the cost of a whole yard.

5. If $\frac{1}{2}$ and $\frac{1}{3}$ of a yard of cloth costs 10 cents, how much does a yard cost?

6. After having sold $\frac{1}{2}$ of a piece of cloth and $\frac{1}{3}$ of it, I had 3 yards remaining. How many yards were there at first?

7. Two boys had each the same number of marbles. One boy bought 2, and they then had 12 in all. How many had each at first?

8. Two boys had each the same number of marbles. One boy sold 3, and they then had 21 in all. How many had each at first?

9. How many apples would be needed to give 8 boys $\frac{3}{4}$ of an apple each?

10. At $\frac{3}{8}$ of a dollar a bushel, how much would 12 bushels of oats cost?

11. A farmer sold $\frac{5}{6}$ of his chickens, and kept 5 for himself. How many had he at first?

12. What is the difference between $\frac{2}{4}$ and $\frac{5}{8}$?

13. What is the difference between $\frac{1}{3}$ and $\frac{2}{9}$?

14. If a boy earns $\frac{7}{8}$ of a dollar a day, and spends $\frac{3}{4}$ of a dollar a day, in how many days will he have saved a dollar?

Reduction of Fractions

1. Change $\frac{2}{3}$ to ninths.

$\frac{2}{3}$ is changed to ninths by multiplying both the numerator and the denominator of $\frac{2}{3}$ by 3. $\frac{2}{3}$ equals $\frac{6}{9}$. In $\frac{6}{9}$ we have three times as many parts as in $\frac{2}{3}$, but these parts are only one third as large.

To change a fraction to larger terms, multiply both the numerator and the denominator by a number which will produce the required terms.

2. Change $\frac{6}{10}$ to fifths.

$\frac{6}{10}$ is changed to fifths by dividing both the numerator and the denominator by 2. $\frac{6}{10}$ equals $\frac{3}{5}$. In $\frac{3}{5}$ we have one half as many parts as in $\frac{6}{10}$, but these parts are twice as large.

To change a fraction to smaller terms, divide the numerator and the denominator by the same number.

Change :

3. $\frac{1}{2}$ to fourths.

4. $\frac{1}{2}$ to eighths.

5. $\frac{2}{3}$ to sixths.

6. $\frac{2}{3}$ to twelfths.

7. $\frac{3}{4}$ to eighths.

8. $\frac{1}{4}$ to twelfths.

9. $\frac{1}{2}$ to sixths.

10. $\frac{1}{3}$ to twelfths.

11. $\frac{1}{4}$ to eighths.

12. $\frac{2}{5}$ to tenths.

13. $\frac{4}{5}$ to tenths.

14. $\frac{5}{6}$ to twelfths.

Change :

15. $\frac{2}{4}$ to halves.

16. $\frac{3}{6}$ to halves.

17. $\frac{6}{9}$ to thirds.

18. $\frac{4}{6}$ to thirds.

19. $\frac{2}{8}$ to fourths.

20. $\frac{6}{8}$ to fourths.

21. $\frac{2}{8}$ to fourths.

22. $\frac{4}{8}$ to halves.

23. $\frac{2}{10}$ to fifths.

24. $\frac{6}{10}$ to fifths.

25. $\frac{8}{12}$ to thirds.

26. $\frac{9}{12}$ to fourths.

40

Squares

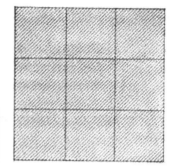

1. How many square inches are there in a square 3 inches long and 3 inches wide?

2. How many square inches in a square 5 inches long?

3. How many square inches in a ten-inch square?

4. How many square inches in a square foot?

5. How many square feet are there in a square yard?

6. Find the number of square inches in 3 square feet.

7. Find the number of square inches in a square yard.

8. How many square feet are there in 5 sq. yd. and 2 sq. feet?

9. How many sq. in. are there in 2 sq. yd., 5 sq. ft., and 10 sq. in.?

10. How many sq. ft. are there in 432 sq. in.?

11. How many sq. yd. are there in 60 sq. ft.?

12. How many 2-in. squares can be made from a 4-in. square?

13. How many 2-in. squares can be made from a 6-in. square?

14. How many 3-in. squares can be made from a 12-in. square?

15. How many squares, 2 feet long, are there in a square floor which is 10 feet long?

41

Compound Quantities

1. Multiply 2 gal. 3 qt. by 3.

2 gal.	3 qt.
	3

6 gal. 9 qt.

8 gal. 1 qt.

First multiply each quantity and then change to larger measures, if possible. As 9 qt. = 2 gal. and 1 qt., we add the 2 gal. to the gallons.

2. Divide 5 bu. 2 pk. 4 qt. by 2.

2)5 bu. 2 pk. 4 qt.

2 bu. 3 pk. 2 qt.

One half of 5 bushels is 2 bushels with 1 bushel over. This 1 bushel with the 2 pecks makes 6 pecks. One half of 6 pecks is 3 pecks; etc.

Multiply :

3.

5 gal. 2 qt. 1 pt.

2

4.

7 bu. 3 pk. 4 qt.

4

5.

10 bu. 2 pk. 7 qt.

5

6.

4 yd. 2 ft. 10 in.

8

Divide :

7.

2)8 bu. 2 pk. 4 qt.

8.

4)9 bu. 1 pk. 4 qt.

9.

2)6 yd. 2 ft. 8 in.

10.

5)8 yd. 2 ft. 3 in.

11. Multiply 12 gal. 3 qt. by 8.

12. Multiply 16 bu. 6 qt. by 7.

13. Divide 14 bu. 6 pk. 4 qt. by 5.

14. Divide 19 yd. 2 ft. 4 in. by 6.

15. Multiply 42 ft. 10 in. by 9.

16. Divide 85 bu. 2 pk. 5 qt. by 4.

42

Similar Fractions

Similar fractions are fractions which have the same denominator.

1. Change $\frac{1}{2}$, $\frac{3}{4}$, and $\frac{5}{8}$ to the same denominator.

$\frac{1}{2} = \frac{4}{8}$ We change $\frac{1}{2}$ and $\frac{3}{4}$ to eighths by multiplying the nu-

$\frac{3}{4} = \frac{6}{8}$ merator and the denominator of $\frac{1}{2}$ by 4, and the numerator

$\frac{5}{8} = \frac{5}{8}$ and the denominator of $\frac{3}{4}$ by 2.

2. Change $\frac{2}{3}$ and $\frac{3}{4}$ to the same denominator.

$\frac{2}{3} = \frac{8}{12}$ We change the fractions to twelfths by multiplying both

$\frac{3}{4} = \frac{9}{12}$ the numerator and the denominator of $\frac{2}{3}$ by 4, and both the numerator and the denominatof of $\frac{3}{4}$ by 3.

Change to the same denominator :

3. $\frac{1}{2}$, $\frac{3}{4}$.

4. $\frac{1}{2}$, $\frac{3}{8}$.

5. $\frac{1}{3}$, $\frac{1}{2}$.

6. $\frac{3}{4}$, $\frac{5}{8}$.

7. $\frac{2}{3}$, $\frac{5}{6}$.

8. $\frac{1}{2}$, $\frac{1}{5}$.

9. $\frac{2}{3}$, $\frac{1}{9}$.

10. $\frac{1}{2}$, $\frac{5}{6}$.

11. $\frac{3}{8}$, $\frac{1}{4}$.

12. $\frac{3}{4}$, $\frac{1}{6}$.

13. $\frac{1}{2}$, $\frac{1}{3}$, $\frac{5}{6}$.

14. $\frac{1}{3}$, $\frac{2}{3}$, $\frac{5}{6}$.

15. $\frac{1}{2}$, $\frac{4}{5}$, $\frac{3}{10}$.

16. $\frac{1}{2}$, $\frac{3}{8}$, $\frac{1}{3}$.

17. $\frac{1}{4}$, $\frac{2}{3}$, $\frac{5}{6}$.

Before fractions can be added or subtracted they must be made similar.

18. Add $\frac{1}{2}$, $\frac{2}{3}$, and $\frac{3}{4}$.

$$\frac{1}{2} = \frac{6}{12}. \quad \frac{2}{3} = \frac{8}{12}. \quad \frac{3}{4} = \frac{9}{12}. \quad \frac{6}{12} + \frac{8}{12} + \frac{9}{12} = \frac{23}{12}.$$

19. From $\frac{4}{5}$ take $\frac{1}{2}$.

$$\frac{4}{5} = \frac{8}{10}. \quad \frac{1}{2} = \frac{5}{10}. \quad \frac{8}{10} - \frac{5}{10} = \frac{3}{10}.$$

Add :

20. $\frac{1}{2}$ and $\frac{5}{8}$.

21. $\frac{2}{3}$ and $\frac{1}{6}$.

22. $\frac{3}{5}$ and $\frac{8}{10}$.

23. $\frac{1}{2} + \frac{1}{3} + \frac{1}{6}$.

24. $\frac{1}{4} + \frac{2}{3} + \frac{6}{12}$.

25. $\frac{5}{6} + \frac{1}{3} + \frac{1}{4}$.

Subtract :

26. $\frac{1}{2}$ from $\frac{9}{10}$.

27. $\frac{3}{4}$ from $\frac{7}{8}$.

28. $\frac{3}{8}$ from $\frac{5}{8}$.

29. $\frac{1}{2} - \frac{1}{8}$.

30. $\frac{3}{4} - \frac{2}{3}$.

31. $\frac{5}{6} - \frac{1}{4}$.

Problems

1. If a quart of oil will last to fill a lamp 4 days, how many days will three gallons last to fill two lamps?

2. If a lamp consumes a quart of oil in 9 hours, how many evenings will a gallon last if a lamp burns 3 hours each evening?

3. How much will 3 gallons and 1 quart of oil cost at 12 cents a gallon?

4. If I feed my chickens 2 quarts of corn a day, how long will a two-bushel bag of corn last?

5. How much will a peck of peanuts cost at the rate of 5 cents a pint?

6. If I pay 10 cents for 4 quarts of apples, how much would a bushel cost at the same rate?

7. If hay sells at 20 dollars a ton, how much do 100 pounds cost?

8. If 100 pounds of coal cost 30 cents, how much will half a ton cost?

9. At 12 cents a foot what will be the cost of ten yards of iron pipe?

10. How many yards is it around a room which is 22 feet long and 17 feet wide?

11. How many square feet are there in one half of the floor of a room 20 feet square?

12. I have 3 pieces of carpet. The first is 5 yards 1 foot long, the second 4 yards 2 feet, and the third $6\frac{1}{2}$ yards long. How much are the 3 pieces together worth at 50 cents a yard?

13. If I step 2 feet at each step, how many times shall I step in walking a mile?

44

Rectangles

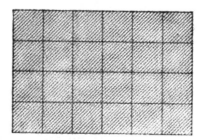

1. What is the area of an oblong 4 inches wide and 6 inches long?

2. What is the area of an oblong 7 feet wide and 4 yards long?

3. What part of a square foot is a rectangle 12 inches long and 6 inches wide?

4. What is the distance around a rectangle 12 feet long and 6 feet wide?

5. How many square feet are there in a rectangle $12\frac{1}{2}$ feet long and 6 feet wide?

6. How many square feet are there in a rectangle 4 yards and 2 feet long and 2 yards and 1 foot wide?

7. How many square inches are there in a pane of glass 12 inches long and 9 inches wide?

8. If there are 4 panes of glass in a window and each pane is 32 in. × 16 in., how many square inches of glass are there in the window?

9. How many square yards of oilcloth will be required to cover a room 6 yards long and $5\frac{1}{2}$ yards wide?

10. A floor is 5 yards square; a second floor is 4 yards square; and a third 3 yards square; how many square yards are there in the three floors?

11. How much will it cost to paint the floor of a room 10 yards long and $7\frac{1}{2}$ yards wide at 12 cents a square yard?

45

Fractions — Drill Work

1. Add $\frac{2}{8}$ and $\frac{3}{8}$.
2. Add $\frac{7}{9}$ and $\frac{1}{9}$.
3. Add $\frac{3}{5}$ and $\frac{9}{10}$.
4. Add $\frac{1}{2}$ and $\frac{2}{3}$.
5. Add $\frac{5}{6}$ and $\frac{2}{3}$.
6. Add $\frac{5}{8}$ and $\frac{1}{4}$.

7. $\frac{1}{3} + \frac{3}{9} + \frac{2}{9} = ?$
8. $\frac{5}{6} + \frac{1}{3} + \frac{1}{2} = ?$
9. $\frac{3}{4} + \frac{5}{8} + \frac{1}{2} = ?$
10. $3\frac{1}{5} + 7 + 4\frac{3}{10} = ?$
11. $5\frac{1}{2} + 6\frac{3}{4} + \frac{7}{8} = ?$
12. $8\frac{5}{9} + 3\frac{2}{3} + 10 = ?$

13. From $\frac{8}{9}$ take $\frac{2}{3}$.
14. From $\frac{4}{5}$ take $\frac{5}{6}$.
15. From $\frac{7}{10}$ take $\frac{2}{5}$.
16. From $\frac{7}{8}$ take $\frac{3}{10}$.
17. From $\frac{5}{8}$ take $\frac{1}{3}$.
18. From $\frac{9}{8}$ take $\frac{3}{4}$.

19. $4\frac{5}{9} - 1\frac{3}{9} = ?$
20. $7\frac{2}{3} - 4\frac{2}{3} = ?$
21. $10\frac{1}{2} - 3\frac{1}{8} = ?$
22. $12\frac{3}{5} - \frac{9}{10} = ?$
23. $6\frac{2}{3} - \frac{5}{6} = ?$
24. $2\frac{7}{10} - 1\frac{4}{5} = ?$

25. Multiply $\frac{2}{3}$ by 9.
26. Multiply $\frac{5}{9}$ by 6.
27. Multiply $\frac{5}{6}$ by 8. •
28. Multiply $\frac{9}{10}$ by 7.
29. Multiply $\frac{7}{8}$ by 5.
30. Multiply $\frac{3}{4}$ by 10.

31. $3\frac{5}{6} \times 4 = ?$
32. $7\frac{1}{5} \times 4 = ?$
33. $2\frac{1}{3} \times 8 = ?$
34. $4\frac{3}{10} \times 6 = ?$
35. $5\frac{3}{4} \times 8 = ?$
36. $8\frac{1}{2} \times 10 = ?$

37. Divide $\frac{8}{9}$ by 2.
38. Divide $\frac{9}{10}$ by 3.
39. Divide $\frac{3}{4}$ by $\frac{3}{8}$.
40. Divide $\frac{4}{3}$ by $\frac{1}{6}$.
41. Divide $\frac{8}{6}$ by $\frac{2}{3}$.
42. Divide $\frac{4}{5}$ by $\frac{3}{10}$.

43. $8\frac{4}{9} \div 4 = ?$
44. $10\frac{5}{6} \div 5 = ?$
45. $8\frac{6}{10} \div 2 = ?$
46. $15\frac{9}{10} \div 3 = ?$
47. $7 \div 2\frac{1}{3} = ?$
48. $15\frac{9}{10} \div 5\frac{3}{10} = ?$

Original Problems

Make problems and solve them:

1. There are 50 pounds in a bushel of meal.

2. A basket of coal weighs 90 lb. There are 2000 lb. in a ton.

3. A grocer divided 12 lb. of tea into 4 oz. packages.

4. Sugar is sold at the rate of 16 lb. for a dollar.

5. A boat sails $5\frac{1}{2}$ miles an hour.

6. A merchant sold a suit of clothes for $\frac{1}{4}$ more than the cost.

7. Two men started at the same place and traveled in opposite directions.

8. A horse travels 9 miles in $\frac{3}{4}$ of an hour.

9. Sugar costs $5\frac{1}{4}$ cents a pound.

10. 10 is $\frac{5}{8}$ of a certain number.

11. A marketman sold $4\frac{3}{8}$ lb. of meat.

12. I bought $5\frac{3}{8}$ lb. of cheese.

13. I sold $\frac{3}{4}$ of a piece of cloth and $\frac{1}{8}$ of it, and had 3 yards left.

14. A room is $14\frac{3}{4}$ ft. long and $12\frac{1}{4}$ ft. wide.

15. A floor is 9 ft. square. It is covered with tiles 6 inches square.

16. A quart of oil lasts to fill a lamp 5 days.

17. 5 chickens will eat 3 quarts of corn in 3 days.

18. Wire netting costs $1\frac{1}{2}$ cents a square foot.

19. 200 pounds of coal cost 50 cents.

20. Hay sells at $22 a ton.

21. A certain window contains 12 panes of glass, each 22 inches long and 18 inches wide.

Surface Measure

1. How many 2 in. squares can be marked off upon a square 10 in. long?

2. How many squares 3 in. long are contained in a square 2 ft. long?

3. How many squares 6 in. long will be contained in a rectangle 5 ft. long and 3 ft. wide?

4. How many tiles 6 in. long and 3 in. wide will cover a square foot of floor?

5. How many tiles 12 in. long and 6 in. wide will be required to lay a floor 15 ft. long and 10 ft. wide?

6. If a room contains 3 windows and each window contains 4 panes of glass 30 in. × 15 in., by how many sq. in. of glass is the room lighted?

7. If a schoolroom is lighted by 4 windows, each containing 12 panes of glass 32 in. × 16 in., by how many sq. ft. of glass is the room lighted?

8. If the above room is 32 ft. long and $30\frac{1}{4}$ ft. wide, how many sq. ft. of floor has it?

9. About how many times as many sq. ft. of floor has the room, as of window glass?

10. It is claimed that a schoolroom should have $\frac{1}{5}$ as much surface of window glass, as of floor. To comply with this rule, with how many more sq. ft. of glass should the above room be provided?

11. In a room 20 ft. long, 15 ft. wide, and 10 ft. high, how many sq. ft. are there in the surface of the wall on one side?

12. How many sq. ft. in the surface of one end of the room?

13. How many sq. yd. are there in the ceiling?

14. How many sq. ft. in the four walls of the room?

48

Add:

1. $75\frac{1}{2}$, $64\frac{3}{4}$, $83\frac{5}{8}$.
2. $46\frac{3}{8}$, $37\frac{1}{6}$, $74\frac{1}{2}$.
3. $92\frac{1}{8}$, $64\frac{4}{9}$, $37\frac{2}{3}$.
4. $64\frac{2}{5}$, $89\frac{1}{2}$, $47\frac{9}{10}$.
5. $57\frac{1}{4}$, $72\frac{1}{3}$, $80\frac{5}{6}$.
6. $65\frac{2}{3}$, $40\frac{9}{10}$, $71\frac{3}{10}$.

7. $258\frac{3}{8} + 762\frac{3}{4}$.
8. $548\frac{2}{8} + 680\frac{5}{6}$.
9. $923\frac{4}{5} + 162\frac{7}{10}$.
10. $201\frac{5}{9} + 837\frac{3}{8}$.
11. $750\frac{2}{3} + 140\frac{11}{12}$.
12. $332\frac{1}{2} + 605\frac{5}{8}$.

Subtract:

13. $45\frac{1}{2}$ from $87\frac{3}{4}$.
14. $38\frac{2}{3}$ from $95\frac{1}{8}$.
15. $72\frac{5}{6}$ from $92\frac{1}{2}$.
16. $51\frac{2}{3}$ from $76\frac{1}{10}$.
17. $42\frac{1}{8}$ from $81\frac{3}{4}$.
18. $84\frac{1}{2}$ from $94\frac{1}{8}$.

19. $752\frac{5}{8} - 629\frac{1}{4}$.
20. $537\frac{1}{2} - 421\frac{1}{6}$.
21. $604\frac{1}{8} - 407\frac{1}{3}$.
22. $582\frac{5}{6} - 290\frac{5}{12}$.
23. $637\frac{1}{2} - 509\frac{9}{10}$.
24. $845\frac{3}{4} - 138\frac{5}{12}$.

Multiply:

25. $275\frac{1}{2}$ by 24.
26. $362\frac{3}{4}$ by 40.
27. $518\frac{3}{8}$ by 56.
28. 420 by $54\frac{5}{6}$.
29. 756 by $84\frac{11}{12}$.
30. 927 by $18\frac{5}{8}$.

31. $242\frac{3}{4} \times 28$.
32. $369\frac{2}{5} \times 80$.
33. $894\frac{2}{3} \times 60$.
34. $246 \times 48\frac{5}{6}$.
35. $480 \times 50\frac{9}{10}$.
36. $608 \times 72\frac{5}{8}$.

Divide:

37. $246\frac{3}{4}$ by 3.
38. $248\frac{4}{5}$ by 4.
39. $750\frac{10}{11}$ by 5.
40. $471\frac{9}{10}$ by 3.
41. $582\frac{6}{7}$ by 6.
42. $394\frac{10}{12}$ by 2.

43. $824\frac{8}{9} \div 4$.
44. $755\frac{5}{8} \div 5$.
45. $369\frac{9}{10} \div 3$.
46. $427\frac{7}{8} \div 7$.
47. $396\frac{6}{7} \div 6$.
48. $850\frac{10}{11} \div 5$.

Square Rods — Acres

1. Find how many square feet there are in a square rod.

2. How many square feet in 8 square rods?

3. How many square rods are there in a lot 10 rods long and 6½ rods wide?

4. Find the number of square feet in a lot 5 rods long and 4 rods wide.

5. What is the value of a house lot which is 6 rods long and 4½ rods wide, at 75 cents a square foot?

6. What is the value of a lot 10 rods long and 8⅙ rods wide, at 90 cents a square foot?

7. Into how many house lots, each 6 rods long and 4 rods wide, can a field be divided which is 30 rods long and 20 rods wide?

8. A lot of land 16 rods long and 10 rods wide would contain exactly an acre. How many square rods are there in an acre?

9. What part of an acre is a lot 10 rods long and 8 rods wide?

10. How many square rods in ¼ of an acre?

11. Into how many lots, each 5 rods long and 4 rods wide, can an acre be divided?

12. Find how many square rods there are in 12¾ acres.

13. If 4 acres were divided into 40 equal parts, how many square rods would each part contain?

14. From a farm containing 125 acres two lots containing 4 acres, 75 square rods and 6 acres, 118 square rods were sold. How much land was there left in the farm?

15. A farmer owning a farm of 80 acres purchased two other lots containing 12 A. 60 sq. rd. and 23 A. 130 sq. rd. What was the size of his farm after purchasing these lots?

Accounts

Add the following and prove the work correct:

1.	2.	3.	4.
$434.27	$879.21	$9861.52	$8364.56
596.80	432.65	975.53	756.43
659.42	71.19	45.28	5421.25
25.25	849.67	75.95	9465.85
969.92	806.57	5673.87	603.08
470.55	72.12	879.47	9200.42
82.21	271.22	17.95	756.55
584.15	952.02	8750.23	59.64
79.73	989.86	785.42	847.36
95.42	78.43	841.81	5200.25
726.71	407.65	2967.63	2121.21
841.63	525.25	742.89	299.89

Find the amount of each of the following accounts:

5. 3 lb. coffee @ 32¢; 10 lb. sugar @ 6¢; 2 qt. milk @ 5¢; 1 gal. molasses, 60¢; 5 lb. raisins @ 8¢; 5 lb. oatmeal @ 4¢.

6. 5 chairs @ $1.50; 2 chamber sets @ $27.50; 28 yd. carpet @ 85¢; 10 window shades @ 35¢; 3 rugs @ $1.50.

7. 50 bu. oats @ 65¢; 10 bu. wheat @ 90¢; 20 bbl. flour @ $4.25; 100 bu. oats @ 32¢; 40 bags corn @ $1.10.

8. 8 doz. eggs @ 22¢; 23 lb. butter @ 15¢; 185 lb. cheese @ 9¢; 22 bu. apples @ 60¢; 30 bu. potatoes @ 35¢.

9. 12 yd. calico @ 6¢; 22 yd. sheeting @ 9¢; 15 yd. gingham @ 11¢; 5 spools thread @ 5¢.

10. 3 shirts @ 45¢; 2 pair cuffs @ 20¢; 3 neckties @ 35¢; 1 hat, $2.00; 4 pair socks @ 23¢.

51

Compound Quantities

1. Multiply 3 gal. 2 qt. 1 pt. by 8.

2. Multiply 28 gal. 3 qt. 1 pt. by 10.

3. Multiply 18 bu. 7 qt. by 12.

4. Multiply 23 yd. 2 ft. 5 in. by 9.

5. Divide 17 gal. 3 qt. 1 pt. by 4.

6. Divide 45 bu. 2 pk. 6 qt. by 7.

7. Divide 74 bu. 3 pk. by 6.

8. Divide 37 ft. 10 in. by 8.

9. Multiply 31 bu. 7 qt. by 25.

10. Multiply 25 ft. 10 in. by 40.

11. Divide 65 gal. 3 qt. by 10.

12. Divide 42 yd. 2 ft. 9 in. by 12.

13. How much milk is there in 12 cans, each containing 3 gal. 3 qt. 1 pt.?

14. A barrel of potatoes contains 2 bu. 3 pk. How many potatoes are there in 17 barrels?

15. 182 bu. 3 pk. 4 qt. of apples are taken to market in 4 equal loads. How many bushels are there in each load?

16. How many feet of iron pipe are there in 12 pieces, each 15 ft. 3 in. long?

17. If it takes 10 flagstones, each 1 yd. 1 ft. 8 in. long, to reach across a street, how wide is the street?

18. It takes 6 boards of equal length to reach across a lot which is 49 ft. 6 in. wide. What is the length of each board?

19. A man bought 10 bu. 3 pk. 4 qt. of beans and put them in 4 boxes which contained equal quantities. What was the quantity in each box?

1. If the product of two numbers is 12 and one of the numbers is 4, what is the other number?

2. If the product of two numbers is 40 and one of the numbers is 8, what is the other number?

3. If the area of a rectangle is 30 sq. ft., and it is 6 ft. long, how wide must it be?

4. If the area of a square is 16 sq. ft., how long is it?

5. What is the length of a square which contains 36 sq. ft.?

6. What is the length of a square containing 49 sq. rods?

7. If a lot which contains 160 sq. rods is 8 rods wide, how long must it be?

8. If a lot which contains half an acre is 10 rods long, what is its width?

9. On a lot which is 20 rods wide, how many rods in length make an acre?

10. How many acres are there in a field 28 rods long and 20 rods wide?

11. How many fence boards 8 ft. 3 in. long will it take to reach around this field?

12. How many fence boards 16 ft. 6 in. long will it take to build a fence around the field three boards high?

13. How much will it cost to mow this field at $1.50 an acre?

14. How much would it cost to plow the field at 5 cents a square rod?

15. How many rods is it around the field?

16. How much hay would it produce at the rate of 3 tons per acre?

Mental Problems

1. What part of 160 is 80?

2. What part of an acre is 40 sq. rods?

3. If I sell $\frac{1}{8}$ of a lot of land and have 20 sq. rods left, how large was the lot?

4. How many ounces are there in $\frac{5}{8}$ of 2 pounds?

5. I paid 50 cents for $\frac{5}{8}$ of a yd. of cloth. How much was that a yd.?

6. If a man can do a piece of work in three days, what part of it can he do in one day?

7. If a man can do a piece of work in five days, what part of it can he do in two days?

8. If a man can do $\frac{4}{5}$ of a piece of work in 4 days, how many days will it take him to do the whole?

9. If two men can do a piece of work in 6 days, how many men would be required to do the work in 3 days?

10. If a family consume $\frac{2}{3}$ of a barrel of flour in 2 months, in how many months will they consume 3 barrels?

11. How many barrels of flour would it take to give 10 persons $\frac{4}{5}$ of a barrel each?

12. If a boy walks $3\frac{1}{4}$ miles an hour, how long will it take him to walk 13 miles?

13. A merchant sold a barrel of flour for $6, which was $\frac{1}{5}$ more than it cost him. What was the cost?

14. If 2 men 44 miles apart travel toward each other, one traveling 6 miles an hour and the other 5 miles an hour, how long will it be before they will meet?

15. How long would it be, if one should travel $2\frac{1}{2}$ miles an hour and the other $1\frac{1}{2}$ miles an hour?

54

Measures

How many square inches are there in:

1. A square 10 inches long?
2. A square 13 inches long?
3. A square 1 ft. 3 in. long?
4. A square 1 ft. 8 in. long?
5. A square 2 ft. 6 in. long?
6. A square 3 ft. 4 in. long?

How many square inches in an oblong:

7. 10 inches long, 9 inches wide?
8. 1 ft. 6 in. long, 11 in. wide?
9. 1 ft. 9 in. long, 1 ft. 4 in. wide?
10. 2 ft. long, 1 ft. 10 in. wide?
11. 3 ft. 2 in. long, 2 ft. 5 in. wide?
12. 3 ft. 8 in. long, 3 ft. wide?

How many times is the one contained in the other:

13. A 3-inch square in a 9-inch square?
14. A square 4 in. long in a square 2 ft. long?
15. A square 3 in. long·in a square 3 ft. long?
16. A square 6 in. long in a square 6 ft. long?
17. A square 2 ft. 4 in. long in a square 7 ft. long?

How many times is the one contained in the other:

18. A 4-inch square in an oblong 1 ft. long, 8 in. wide?
19. A square 6 in. long in an oblong 10 ft. × 6 ft.?
20. A square 8 in. long in an oblong 12 ft. × 4 ft. 8 in.?
21. An oblong 2 ft. × 1 ft. 3 in. in an oblong 10 ft. × 2 ft. 6 in.?
22. An oblong 2 ft. 6 in. × 1 ft. in an oblong 12 ft. 6 in. × 4 ft.?

55

Acres — Square Miles

First estimate the results mentally.

1. How many acres are there in a field 64 rods long and 10 rods wide?

2. How many acres are there in a farm consisting of 2 fields, one of which is 80¼ rods long and 40 rods wide, and the other 100 rods long and 32⅝ rods wide?

3. Find how many acres there are in a sq. mile.

4. How many acres are there in a field ¼ of a mile square?

5. If a sq. mile of land were divided into 16 farms, how many acres would each farm contain?

6. How many sq. miles are there in a town which is 6 miles long and 5½ miles wide?

7. Find how many acres there are in a town which is 4 miles long and 3 miles wide.

8. If a town which contains 33 sq. miles is 6 miles long, how wide must it be?

9. How many farms of 160 acres each are contained in a sq. mile?

10. How many farms of 160 acres each would be contained in the state of Rhode Island, which has an area of 1250 sq. miles?

11. The area of the state of New York is 49,170 sq. miles. How many states of the size of Rhode Island would the state of New York make?

12. The area of Alaska is 590,884 sq. miles. How many states of the size of New York would Alaska make?

13. How many states of the size of Rhode Island would Alaska make?

1. How much change shall I receive if I buy 4 five-cent stamps, 20 two-cent stamps, and 15 1-cent stamps, and pay with a dollar bill?

2. If I buy half a doz. oranges @ 25¢ a doz., 2 qt. apples @ 5¢ a qt., and half a pound of candy @ 35¢ a pound, and give a half dollar in payment, how much change shall I receive?

3. How much change shall I receive, if I buy 10 lb. sugar @ 5¢, 1 lb. butter @ 28¢, and 2 doz. eggs @ 24¢, and pay with a two-dollar bill?

4. If I purchase at the store 2 lb. tea @ 65¢, 2 gal. molasses @ 55¢, 1½ lb. butter @ 25¢, and 3 doz. oranges at the rate of 2 doz. for a quarter, what will be the amount of my bill?

5. Find the amount of the following items : 10 bbl. flour @ $5.50 ; 200 lb. sugar @ 4⅚¢ ; 50 bu. corn @ 97¢ ; 75 bu. wheat @ 90¢ ; 100 lb. raisins @ 7¼¢.

6. A man deposited $2500 in a bank and gave checks drawing out the following sums ; $250, $75.84, $125.83, $360.49, $495, $12.45. How much had he left in the bank?

7. I have deposited in the savings bank the following sums at various times : $150, $75, $40, $60. The following items of interest have been added : $4.23, $3.36, $2.78, $4.15. I have withdrawn the following sums : $20, $35, $60. With how much am I now credited at the bank?

8. If a man's monthly salary is $100, and he expends during the month $20 for rent, $28.50 for food, $18.25 for clothing, $4.25 for fuel, $1.25 for gas, $3 for medical service, $6.50 for car fare, and $4.75 for other incidentals, how much does he save?

Miscellaneous Problems

1. Add 4 gal. 3 qt., 5 gal. 2 qt., and 4 gal. 1 qt.

2. From 8 bu. 2 pk. 7 qt. take 4 bu. 3 pk. 5 qt.

3. In each of five boxes there are 10 bu. 3 pk. 6 qt. of potatoes. How many are there in all?

4. If I should have in my store 13 bu. 3 pk. 4 qt. of apples and should sell each customer a peck and a half, how many customers would they supply?

5. If it takes 5½ yards of cloth to make a coat, how many coats can be made from 27½ yards?

6. If it takes 10¾ yards to make a dress, how many yards will it take to make 8 dresses?

7. A man contracts to build a sewer for $1.75 a foot. How much does it cost the city if the sewer is 10 rods 7 feet long?

8. If a contractor builds a sewer for $1.50 a foot, and the materials and labor cost him $1.23 a foot, how much will he make in building a sewer 22 rd. 4 yd. 2 ft. long?

9. Find the number of sq. yd. in a pavement 24 ft. wide and 150 ft. long.

10. How many sq. yd. of cloth will it take to cover the top of a desk which is 3 ft. wide and 4 ft. 6 in. long?

11. How many sq. ft. of boards will it take to build a tight board fence 4½ feet high around a lot 150 feet long and 100 feet wide?

12. How many posts set 8 feet apart will be required to build a fence 48 feet long?

13. How many posts set 10 feet apart will it take to build a fence 20 rods long?

Drill Work

How many square inches in :

1. A square 8 inches long?
2. A square 12 inches long?
3. A square 18 inches long?
4. A square 2 feet long?
5. A square 1 yard long?
6. A square 7 feet long?

How many square inches in a rectangle :

7. 9 in. long, 5 in. wide?
8. 10 in. long, 8 in. wide?
9. 18 in. long, 15 in. wide?
10. 2 ft. long, 16 in. wide?
11. 1 yd. long, 1 ft. wide?
12. 2 yd. long, 2 ft. wide?

Add :

13. 9 gal. 3 qt. 1 pt. ; 7 gal. 2 qt. ; 10 gal. 1 qt. 1 pt.
14. 7 bu. 3 pk. 7 qt. ; 4 bu. 1 pk. 5 qt.; 3 pk. 6 qt.
15. 9 yd. 2 ft. 10 in. ; 1 ft. 11 in. ; 16 yd. 8 in.

Subtract :

16. 4 gal. 3 qt. 1 pt. from 12 gal. 1 qt. 1 pt.
17. 10 bu. 1 pk. 7 qt. from 20 bu. 3 pk. 2 qt.
18. 25 yd. 2 ft. 10 in. from 40 yd. 1 ft. 5 in.

Multiply :

19. 12 gal. 3 qt. 1 pt. by 5.
20. 75 bu. 3 pk. 5 qt. 1 pt. by 8.
21. 10 yd. 2 ft. 9 in. by 40.

Divide :

22. 9 gal. 3 qt. 1 pt. by 3.
23. 16 bu. 3 pk. 4 qt. by 5.
24. 85 yd. 1 ft. 7 in. by 9.

Original Problems

Make problems and solve them:

1. A floor 12 ft. square is covered with tiles 4 inches wide and 3 inches long.

2. A certain schoolroom has 6 windows, each containing 12 panes of glass.

3. 12 acres of land are divided into 20 equal parts.

4. A certain room is 20 ft. long, 15 ft. wide, and 10 ft. high.

5. A man owning 2 acres of land divided it into lots 10 rods long and 4 rods wide.

6. The land on a certain street costs 50 cents a square foot.

7. The product of two numbers is 72.

8. The area of a rectangle is 42 sq. ft.

9. A lot containing half an acre is 8 rods long.

10. A farmer has to build a fence around a field which is 40 rods long and 20 rods wide.

11. It will cost $3 an acre to plow this field.

12. A man can do a piece of work in 6 days.

13. A boy walked 16 miles.

14. A number increased by $\frac{1}{8}$ of itself equals 12.

15. $\frac{3}{5}$ of the pupils in a certain room are girls, and there are 12 boys.

16. A field which is 32 rods long contains 3 acres.

17. A certain town is 5 miles long and 3 miles wide.

18. A certain town is 6 miles square.

19. It takes $8\frac{1}{4}$ yards of cloth to make a dress.

20. A farmer built a fence of posts and boards around a field 20 rods long and 10 rods wide.

60

$$\text{▨▨▨} + \text{▨▨} = 9$$

1. A number increased by $\frac{1}{2}$ of itself equals 9. What is the number?

A number increased by $\frac{1}{2}$ of itself equals $\frac{2}{2} + \frac{1}{2}$ or $\frac{3}{2}$ of itself. 9 is $\frac{3}{2}$ of what number?

2. What number increased by $\frac{1}{3}$ of itself equals 8?

3. What number is that which when diminished by $\frac{1}{3}$ of itself equals 10?

A number diminished by $\frac{1}{3}$ of itself equals $\frac{3}{3} - \frac{1}{3}$ or $\frac{2}{3}$ of itself.

4. $2\frac{1}{2}$ times a certain number equals 5. What is the number?

5. $3\frac{1}{2}$ times a certain number equals 7. What is the number?

6. A boy lost $\frac{2}{3}$ of his money and had 10 cents left. How much had he at first?

7. $\frac{5}{8}$ of the pupils in a certain school are girls, and there are 16 boys. How many girls are there?

8. If $3\frac{1}{2}$ pounds of sugar cost 21 cents, how much does a pound cost?

9. A boy gave his sister 5 oranges, kept 3 himself, and divided the remainder equally among 6 children, who received 4 oranges each. How many had he at first?

10. If 2 men 80 miles apart travel toward each other, one traveling 3 miles an hour and the other 5 miles an hour, how far apart will they be at the end of four hours?

11. If 2 men start at the same point and travel in opposite directions, one traveling $3\frac{1}{4}$ miles an hour and the other $2\frac{1}{2}$ miles an hour, how far apart will they be at the end of 6 hours?

Townships

N.

6	5	4	3	2	1
7	8	9	10	11	12
18	17	16	15	14	13
19	20	21	22	23	24
30	29	28	27	26	25
31	32	33	34	35	36

W.   E.

S.

The western states are divided into townships six miles square. Each township is divided into 36 sections as represented in the diagram.

1. How long is each section?

2. How many acres are there in each section?

The sections are frequently divided into half sections, quarter sections, or smaller parts as convenient for farms.

3. How many rods long is a quarter section?

4. How many square rods are there in a half section?

5. In a farm described as the southwest quarter of section 28, how many acres are there?

6. How many rods is it around this farm?

7. How much would it cost to build a fence around it at 60 cents a rod?

8. How many feet is it directly across the farm?

9. If a man should step $2\frac{1}{2}$ feet at a step, and should take 100 steps a minute, how long would it take him to walk across the farm?

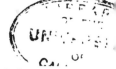

1. How much is one third and one fifteenth? fifth and one fifteenth? One third less one fifth? One third less two fifteenths?

2. How much is two thirds and one fifteenth? One third and four fifteenths? Two thirds less seven fifteenths? Fourteen fifteenths less two fifths?

3. How much is one half and one fourteenth? One seventh and one fourteenth? One half and one seventh? One seventh less one fourteenth?

4. How much is one half and three fourteenths? Five sevenths and one fourteenth? Five fourteenths less two sevenths? Thirteen fourteenths less one half?

5. How much is two thirds and three fifteenths? Six sevenths less one half? Four fifths and two thirds? Three fifths less one third?

6. $\frac{1}{3} + \frac{1}{5} + \frac{1}{15}$

7. $\frac{2}{3} + \frac{2}{15} + \frac{4}{5}$

8. $\frac{2}{3} + \frac{4}{5} - \frac{4}{15}$

9. $\frac{1}{2} + \frac{1}{7} + \frac{1}{14}$

10. $\frac{2}{7} + \frac{1}{2} + \frac{5}{14}$

11. $\frac{3}{2} + \frac{9}{14} + \frac{5}{7}$

12. $\frac{4}{5} + \frac{4}{15} - \frac{1}{5}$

13. $\frac{9}{14} - \frac{1}{2} + \frac{1}{7}$

14. $\frac{14}{15} - \frac{1}{3} + \frac{3}{5}$

Add:

15.	16.	17.	18.	19.
$5\frac{1}{3}$	$8\frac{2}{3}$	$6\frac{1}{2}$	$24\frac{6}{7}$	$95\frac{3}{5}$
$6\frac{2}{5}$	$7\frac{4}{5}$	$4\frac{4}{7}$	$36\frac{1}{2}$	$68\frac{2}{5}$
$8\frac{2}{15}$	$9\frac{4}{5}$	$9\frac{3}{14}$	$49\frac{5}{14}$	$85\frac{4}{15}$
$9\frac{3}{5}$	$6\frac{2}{3}$	$2\frac{5}{7}$	$82\frac{3}{7}$	$74\frac{4}{5}$

Subtract:

20.	21.	22.	23.	24.
$9\frac{1}{3}$	$10\frac{2}{3}$	$12\frac{2}{7}$	$24\frac{1}{2}$	$42\frac{1}{5}$
$1\frac{2}{15}$	$5\frac{2}{5}$	$3\frac{3}{14}$	$12\frac{2}{7}$	$16\frac{1}{3}$

Time Accounts

1. If it is understood in a mill that ten hours constitute a day's work, what will be the week's wages of a man who works 55 hours during the week and receives wages at the rate of $1.25 per day?

2. What will be the week's wages of a man who gets $2.00 a day, and works on Monday, $9\frac{1}{2}$ hours; Tuesday, 10 hours; Wednesday, $8\frac{1}{2}$ hours; Thursday, $10\frac{1}{4}$ hours; Friday, $9\frac{1}{4}$ hours; and Saturday, $7\frac{1}{2}$ hours?

The following is a section from the time sheet of a mill.

	MON. hrs.	TUES. hrs.	WED. hrs.	THURS. hrs.	FRI. hrs.	SAT. hrs.	WAGES.
A	10	$9\frac{1}{4}$	$9\frac{1}{2}$	8	9	$10\frac{1}{4}$	$1.75
B	$9\frac{3}{4}$	$10\frac{1}{2}$	10	$10\frac{3}{4}$	11	$10\frac{1}{2}$	1.80
C	9	$8\frac{1}{2}$	$8\frac{3}{4}$	10	$9\frac{1}{4}$	$8\frac{1}{2}$	2.20
D	10	$7\frac{1}{2}$	0	0	$9\frac{1}{2}$	8	1.20
E	$9\frac{1}{2}$	10	9	$8\frac{1}{4}$	$9\frac{1}{4}$	10	2.00

3. Reckon the week's wages of A.

4. How many hours overtime did B work during the week?

5. If C has a wife and three children, how much did his week's wages yield for the support of each member of his family?

6. How much money did D lose by being absent Wednesday and Thursday?

7. How much more did E earn than A?

8. The total time of the five men for the week equals how many full days' work of 10 hours each?

9. Find the total week's wages of the five men.

64

Fractions — Drill Work

Add:

1. $24\frac{5}{8} + 37\frac{3}{4} + 42\frac{1}{2}$
2. $32\frac{2}{3} + 46\frac{5}{8} + 83\frac{1}{3}$
3. $70\frac{1}{2} + 56\frac{2}{3} + 65\frac{5}{6}$
4. $91\frac{3}{4} + 72\frac{1}{3} + 29\frac{1}{6}$
5. $57\frac{2}{5} + 62\frac{1}{2} + 80\frac{9}{10}$
6. $47\frac{6}{7} + 52\frac{11}{14} + 65\frac{7}{8}$
7. $75\frac{2}{3} + 64\frac{3}{5} + 95\frac{7}{15}$
8. $38\frac{1}{2} + 93\frac{3}{4} + 50\frac{5}{6}$
9. $853\frac{5}{8} + 148\frac{3}{4}$
10. $638\frac{4}{5} + 783\frac{6}{10}$
11. $921\frac{2}{3} + 345\frac{5}{6}$
12. $718\frac{6}{7} + 327\frac{5}{14}$
13. $802\frac{2}{3} + 737\frac{3}{4}$
14. $560\frac{2}{3} + 315\frac{1}{10}$
15. $837\frac{1}{3} + 365\frac{4}{5}$
16. $548\frac{2}{3} + 716\frac{5}{9}$

Subtract:

17. $56\frac{7}{8} - 37\frac{3}{4}$
18. $84\frac{2}{3} - 71\frac{1}{4}$
19. $93\frac{1}{2} - 25\frac{1}{3}$
20. $75\frac{3}{7} - 50\frac{3}{14}$
21. $68\frac{3}{4} - 21\frac{3}{4}$
22. $58\frac{1}{3} - 18\frac{2}{15}$
23. $45\frac{1}{8} - 27\frac{1}{4}$
24. $69\frac{2}{3} - 29\frac{5}{6}$
25. $120 - 82\frac{1}{2}$
26. $265 - 58\frac{5}{6}$
27. $830 - 125\frac{2}{7}$
28. $627\frac{4}{5} - 100\frac{3}{10}$
29. $365\frac{2}{3} - 180\frac{2}{5}$
30. $720\frac{1}{4} - 120\frac{1}{8}$
31. $372\frac{2}{3} - 175\frac{5}{9}$.
32. $541\frac{1}{7} - 231\frac{3}{14}$

Multiply:

33. $25\frac{2}{3} \times 15$
34. $38\frac{3}{4} \times 24$
35. $55\frac{5}{6} \times 72$
36. $67\frac{3}{7} \times 56$
37. $75 \times 39\frac{4}{5}$
38. $81 \times 42\frac{5}{9}$
39. $96 \times 73\frac{11}{12}$
40. $90 \times 36\frac{14}{15}$
41. $135\frac{3}{4} \times 200$
42. $237\frac{5}{8} \times 400$
43. $562\frac{5}{6} \times 360$
44. $495\frac{3}{7} \times 140$
45. $730\frac{8}{9} \times 639$
46. $316\frac{2}{3} \times 480$
47. $524\frac{1}{12} \times 852$
48. $237\frac{14}{15} \times 930$

Measure of Lumber

A single foot of lumber would be a piece of board one foot square and one inch thick. Boards are regarded as one inch thick unless it is otherwise specified.

1. How many feet of lumber are there in a board 3 ft. long and 1 ft. wide?

2. How many feet in a board 2 ft. wide and 10 ft. long?

3. How many feet in a board $1\frac{1}{2}$ ft. wide and 12 ft. long?

4. How many feet in a piece of plank 1 ft. square and 2 in. thick?

5. How many feet in a plank 1 ft. wide, 2 in. thick, and 8 ft. long?

Since there would be 8 board feet in the plank if it were 1 in. thick, there are 16 board feet in the plank 2 in. thick.

6. How many feet in a plank 16 ft. long, $2\frac{1}{2}$ ft. wide, and 3 in. thick?

7. How many feet in a joist 6 in. wide, 4 in. thick, and 10 ft. long?

8. How many feet in a joist 4 in. × 3 in. and 15 ft. long?

9. How many feet in a plank 15 in. wide, 16 ft. long, and $2\frac{1}{2}$ in. thick?

10. How many feet in 10 pieces of timber each 12 ft. long, 9 in. wide, and 6 in. thick?

11. How many feet of boards will it take for the floor of a room which is 22 ft. long and 14 ft. wide, allowing 20 sq. ft. for waste in fitting?

12. How many feet of boards will it take to board the side of a house which is 50 ft. long and 26 ft. high, if the space left for the windows will equal the waste in fitting?

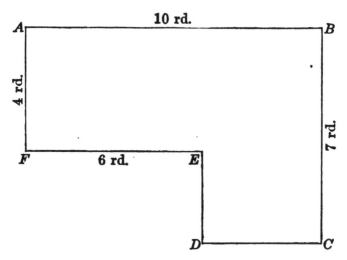

1. How many rods is it around this lot?

2. What is the distance in feet from *E* to *D*?

3. What is the distance in yards from *E* to *C*, measuring around by way of *D*?

4. Which is the nearer way from *A* to *C*, by way of *B* or by way of *FED*; and how much?

5. If a man should step 3 feet at each step, how many steps would he take in following the border of the lot from *F* to *C*?

6. How many boards each 8¼ feet long would it take to go around the entire lot?

7. If the fence posts were set half a rod apart, how many posts would have to be set to build a fence across the side *AB*?

8. How many feet of boards would it take to nail a board to the tops of the posts entirely around the lot, if the boards were 6 in. wide?

9. How many feet of wire would it take to fasten two wires to the posts beneath the boards entirely around the lot?

Averages

1. What number is as much larger than 8 as it is smaller than 14?

2. What is the average length of three boards which are 4 ft. long, 5 ft. long, and 6 ft. long?

The total length of the three boards is 4 ft. + 5 ft. + 6 ft. or 15 ft. The average length is 15 ft. ÷ 3, or 5 ft.

3. If a boy rides a bicycle 6 miles the first hour, 8 miles the second, 8 miles the third, and 10 miles the fourth, what is his average speed?

The total number of miles must be divided by 4. Why?

4. What is the average of the numbers 6, 8, 10, and 12?

5. Find the average of the numbers 5, 9, 7, 12, 10, 6, and 14.

By what number should this total be divided?

6. What is the average of the numbers 123, 245, 334, and 580?

7. If the attendance of pupils at a school is for Monday, 40, Tuesday, 38, Wednesday, 41, Thursday, 35, and Friday, 38, what is the average attendance for the week?

8. If there were three instances of tardiness on Monday, seven on Tuesday, four on Wednesday, five on Thursday, and six on Friday, what was the average of tardiness for each half day?

9. If the weekly average marks of a pupil are $4\frac{3}{4}$, $3\frac{1}{2}$, $4\frac{1}{2}$, 5, $3\frac{3}{4}$, $2\frac{1}{2}$, 4, $3\frac{1}{4}$, and $4\frac{3}{4}$, what is his average for the quarter?

10. If there were 5 rain storms in June, during which the amounts of rainfall were $1\frac{2}{8}$ in., $\frac{4}{8}$ in., $1\frac{1}{8}$ in., $\frac{2}{8}$ in., and $1\frac{3}{10}$ in., what was the total rainfall for the month and the average for each storm?

Time

1. How many minutes are there in 1 hour and 45 minutes?

2. How many seconds in 1 hour and 20 minutes?

3. How many days and hours are there in 3000 minutes?

4. Add the following periods of time: 6 hr. 30 min. 10 sec.; 5 hr. 42 min. 35 sec.; 14 hr. 5 min. 25 sec.

5. Multiply 4 hr. 18 min. 12 sec. by 15.

6. If 13 hr. 17 min. were divided into 4 equal periods, how long would each period be?

7. How far will a man walk in two weeks if he walks 25 miles a day and rests on Sunday?

8. If a train goes at the rate of a mile in 2 minutes, how long will it take it to go 150 miles?

9. If a man sleeps 6 hours each night, what part of the time does he sleep?

10. How many hours does a man work in a year if he works 10 hours a day and has 2 weeks of vacation?

11. If a train runs at the rate of 2 miles in 3 minutes, how far will it run in 30 minutes?

30 is how many times 3? How many times 2 miles will the train run in the 30 minutes?

12. If a train runs at the rate of 4 miles in 5 minutes, how long will it take it to run from Providence to Boston, a distance of 44 miles?

13. If a train runs at the rate of 40 miles an hour, but loses 5 minutes each time it stops, how long will it take it to run 160 miles and make four stops on the way?

14. If a cigar smoker spends 20 ¢ each day for cigars, how much will the expense amount to in 10 years, if 3 of the years contain 366 days each?

Time between Dates

1. Find the exact number of days from May 20 to Aug. 5.

11
30
31
5
——
77

Write in a column the number of days in each of the months and add them. In May there are 11 days remaining. In June there are 30 days, in July 31 days, and from July 31 to Aug. 5, 5 days. In all there are 77 days.

2. Find the number of months and days from Sept. 8 to Dec. 19.

Find the whole number of months and then reckon the days remaining. From Sept. 8 to Dec. 8 there are 3 months, and from Dec. 8 to Dec. 19 there are 11 days.

3. Give the number of days in each month of the year.

4. How many days are there from Aug. 1 to Aug. 8? From Aug. 1 to Aug. 31? From Aug. 30 to Sept. 2?

5. Find the exact number of days from Jan. 10, 1899, to Mar. 15, of the same year.

6. Find the exact number of days from May 1 to Sept. 30.

7. How many months are there from Jan. 1 to July 1?

8. How many months from Mar. 10 to Nov. 10?

9. How many even months and how many days over from June 1 to Aug. 5?

10. How many months and how many days over from Aug. 7 to Dec. 10?

How many months and days over from:

11. Jan. 10 to May 21?	**16.** July 25 to Oct. 10?
12. Mar. 9 to May 30?	**17.** July 12 to Oct. 1?
13. Feb. 14 to Apr. 1?	**18.** Aug. 31 to Nov. 15?
14. June 1 to Sept. 1?	**19.** Sept. 10 to Jan. 31?
15. June 1 to Aug. 20?	**20.** Nov. 19 to Feb. 10?

70

Angles

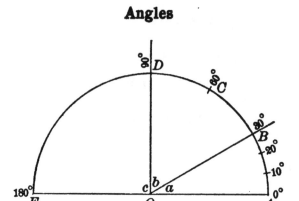

An **angle** is the measure of the difference in direction of two lines.

In the figure the angle at a is formed by the lines OA and OB. The size of an angle is determined by forming a circle with its center at the vertex of the angle and dividing the circumference into 360 equal parts. Since the angle a includes between its sides 30 of the 360 parts of the circle, it is an angle of 30 degrees.

1. If the line OB were revolved far enough to reach the point C, how large would the angle a then be?

2. The angle AOD is called a right angle. A right angle is an angle of how many degrees?

Right angle

Acute angle

Obtuse angle

3. How large is the angle b, which is the difference in direction of the lines OB and OD?

An acute angle is smaller than a right angle. An obtuse angle is larger than a right angle.

4. If an angle of 40° is added to an angle of 45°, will the two make an obtuse angle or an acute angle?

71

Miscellaneous Drill Work

1. Add $\frac{3}{7}$ and $\frac{5}{14}$.
2. Add $\frac{1}{2}$ and $\frac{6}{7}$.
3. Add $\frac{7}{10}$ and $\frac{4}{5}$.
4. Add $\frac{1}{3}$ and $1\frac{1}{6}$.
5. Add $\frac{2}{3}$ and $\frac{1}{6}$.

6. $\frac{5}{7} + \frac{9}{14} + \frac{1}{7} = ?$
7. $\frac{1}{2} + \frac{1}{7} + \frac{1}{14} = ?$
8. $\frac{3}{5} + \frac{1}{2} + \frac{9}{10} = ?$
9. $\frac{1}{3} + \frac{2}{15} + \frac{2}{3} = ?$
10. $\frac{1}{5} + \frac{2}{3} + \frac{4}{5} = ?$

Find the number of feet of lumber in a board :

11. 10 ft. long, 1 ft. wide.
12. 10 ft. long, 2 ft. wide.
13. 8 ft. long, $1\frac{1}{2}$ ft. wide.
14. $6\frac{1}{2}$ ft. long, 1 ft. wide.
15. 10 ft. long, 6 in. wide.
16. 12 ft. long, 1 ft. wide.
17. 12 ft. long, 18 in. wide.
18. 7 ft. 6 in. long, 2 ft. wide.

Find the average of :

19. 5, 7, 8, 9, 11.
20. 6, 9, 13, 17, 20.
21. 4, 7, 19, 42.
22. 84, 71, 60, 27, 16.
23. 112, 137, 116, 153.
24. 250, 221, 193, 577.

25. $1\frac{3}{4}, \frac{3}{4}, \frac{7}{8}, 1\frac{1}{8}$.
26. $\frac{4}{5}, \frac{1}{10}, \frac{1}{2}, 1\frac{1}{5}, 2\frac{9}{10}$.
27. 12 qt., 17 qt., 10 qt.
28. 15 bu. 3 pk., 12 bu. 2 pk.
29. $2\frac{1}{4}$ days, $4\frac{1}{2}$ days, 3 days.
30. 10 yd. $2\frac{1}{4}$ ft., 21 yd. $1\frac{1}{4}$ ft.

Find the number of months and days from :

31. Jan. 1 to Feb. 12.
32. Mar. 5 to May 5.
33. Mar. 5 to May 16.
34. Apr. 30 to June 1.
35. Apr. 30 to June 28.

36. Aug. 1 to Aug. 31.
37. June 28 to Aug. 1.
38. Sept. 15 to Oct. 15.
39. Sept. 15 to Nov. 15.
40. Dec. 10 to Jan. 18.

Original Problems

Make problems and solve them:

1. A township is six miles square.

2. Section 24 of a certain township is divided into 4 farms.

3. A family took 3 pints of milk a day from the milkman for the month of June.

4. A merchant sold goods for $\frac{7}{8}$ of their cost.

5. A merchant was obliged to sell a lot of flour for $\frac{4}{5}$ of the cost.

6. I sold cloth at a profit of 10 cents a yard.

7. A man left $\frac{1}{3}$ of his property to his wife, $\frac{1}{6}$ to his daughter, and the remainder to his son.

8. 3 men can do $\frac{2}{5}$ of a piece of work in a day.

9. In a shop where 10 hours make a day's work, a man who has $2 a day worked $52\frac{1}{2}$ hours during the week.

10. A board is 9 inches wide and 12 feet long.

11. A lot is 16 rods long and 14 rods wide. The wire for the fence costs 4 cents a rod for one wire.

12. The daily attendance at school for the week was 35, 38, 32, 37, and 30.

13. A man walked 15 miles a day.

14. A train runs at the rate of 3 miles in 5 minutes.

15. A boy begins at his 12th birthday to save ten cents a week.

16. The summer vacation continued from June 30th till Sept. 12th.

17. At a certain time of the year the days are 12 hours, 25 minutes long.

18. Feb. 1, 1900, was Thursday.

73

Review Problems

1. What is the width of a door, the surface of one side of which is 28 square feet and whose length is 8 feet?

2. If a certain field contains exactly 2 acres and is 32 rods long, how wide is it?

3. The area of the state of Delaware is 2050 square miles. What is its area in acres?

4. The population of Delaware in 1900 was 184,735. How many acres did the state average for each person?

5. How many feet of boards will be required to make a board walk 4 ft. 4 in. wide and 90 ft. long?

6. How many thousand feet of lumber will it take to build a tight board fence 10 feet high and 20 rods long?

7. At $22 a thousand feet, what will be the cost for boards to cover the two sides of the roof of a house, each of which is 60 feet long and 35 feet wide?

8. The ages of four children are 12 yr. 3 mo., 10 yr. 6 mo., 13 yr. 9 mo., and 12 yr. 10 mo. Find the average of their ages.

9. For a week in January at 8 o'clock in the morning the thermometer indicated as follows: Sunday 8°, Monday 13°, Tuesday 5°, Wednesday 9°, Thursday 15°, Friday 14°, and Saturday 20°. What was the average for the week?

10. President Lincoln was born Feb. 12, 1809, and died Apr. 15, 1865. At what age did he die?

11. Find the exact number of days from Jan. 12, 1896, to Nov. 18 of the same year.

12. If from a right angle an angle of 20° and an angle of 30° are taken, how large is the angle which remains?

13. If ten equal angles are arranged around one point so as to occupy an entire circle, how large is each angle?

74

Compound Quantities

Change :

1. 8 gal. 3 qt. to pints.
2. 24 bu. 3 pk. 5 qt. to quarts.
3. 16 yd. 8 in. to inches.
4. 4 rd. 13 ft. 4 in. to inches.
5. 27 sq. ft. 35 sq. in. to square inches.
6. 5 sq. rd. 100 sq. ft. to square feet.

Change:

7. 67 pt. to gallons, quarts, and pints.
8. 75 qt. to bushels, pecks, and quarts.
9. 207 in. to yards, feet, and inches.
10. 210 in. to rods, feet, and inches.
11. 300 sq. in. to square feet and square inches.
12. 320 sq. ft. to square rods and square feet.

Add :

13. 12 gal. 2 qt. 1 pt., 9 gal. 3 qt.
14. 45 bu. 2 pk. 7 qt., 37 bu. 3 pk. 5 qt.
15. 25 yd. 2 ft. 5 in., 13 yd. 1 ft. 11 in.
16. 12 rd. 15 ft. 9 in., 17 rd. 14 ft. 10 in.
17. 89 sq. ft. 95 sq. in., 4 sq. rd. 15 sq. ft. 25 sq. in.
18. 200 A. 100 sq. rd., 180 A. 150 sq. rd.

Subtract :

19. 16 bu. 5 qt. from 28 bu. 3 pk. 4 qt.
20. 7 yd. 1 ft. 7 in. from 20 yd. 2 ft. 5 in.
21. 21 rd. 12 ft. from 52 rd. 10 ft. 8 in.
22. 25 sq. yd. 7 sq. ft. from 29 sq. yd. 4 sq. ft.
23. 116 sq. rd. 150 sq. ft. from 182 sq. rd. 100 sq. ft.
24. 75 A. 92 sq. rd. from 250 A. 80 sq. rd.

Improper Fractions — Mixed Numbers

A **proper fraction** is less than a whole number. When written, its numerator is less than its denominator.

An **improper fraction** is either equal to a whole number or greater than a whole number. When written, its numerator is either equal to its denominator or greater than its denominator. It may either indicate fractional parts, or simply that the number above the line is to be divided by the number below the line. $\frac{3}{4}$ is a proper fraction. $\frac{11}{8}$ is an improper fraction.

A **mixed number** is a whole number with a fraction. $12\frac{1}{8}$ is a mixed number.

1. Change $8\frac{2}{5}$ to an improper fraction.

In 1 whole there are 5 fifths. In 8 wholes there are 40 fifths. $\frac{40}{5}$ and $\frac{2}{5}$ are $\frac{42}{5}$.

2. Change $\frac{74}{9}$ to a mixed number.

9 ninths make 1 whole. In 74 ninths there are 8 wholes and 2 ninths over, or $8\frac{2}{9}$.

Change to improper fractions :

3. $9\frac{2}{3}$.	**8.** $28\frac{1}{5}$.	**13.** $62\frac{5}{8}$.	**18.** $37\frac{10}{11}$.
4. $8\frac{9}{10}$.	**9.** $32\frac{5}{6}$.	**14.** $21\frac{9}{13}$.	**19.** $64\frac{7}{12}$.
5. $12\frac{3}{8}$.	**10.** $49\frac{11}{12}$.	**15.** $47\frac{1}{12}$.	**20.** $95\frac{1}{15}$.
6. $14\frac{6}{7}$.	**11.** $73\frac{5}{11}$.	**16.** $79\frac{5}{8}$.	**21.** $122\frac{9}{10}$.
7. $25\frac{9}{10}$.	**12.** $64\frac{7}{10}$.	**17.** $96\frac{1}{14}$.	**22.** $256\frac{5}{12}$.

Change to mixed numbers :

23. $\frac{25}{4}$.	**28.** $\frac{45}{10}$.	**33.** $\frac{115}{12}$.	**38.** $\frac{285}{10}$.
24. $\frac{38}{5}$.	**29.** $\frac{87}{12}$.	**34.** $\frac{187}{13}$.	**39.** $\frac{264}{11}$.
25. $\frac{64}{7}$.	**30.** $\frac{79}{9}$.	**35.** $\frac{146}{14}$.	**40.** $\frac{318}{14}$.
26. $\frac{73}{10}$.	**31.** $\frac{92}{8}$.	**36.** $\frac{213}{11}$.	**41.** $\frac{321}{12}$.
27. $\frac{88}{11}$.	**32.** $\frac{87}{7}$.	**37.** $\frac{250}{9}$.	**42.** $\frac{430}{15}$.

Problems from Geography

1. The state of Indiana contains 36,350 square miles. It is about 250 miles long. About how wide must it be?

2. The state of Ohio is about 200 miles long. It contains 41,060 square miles. About how far is it across the state?

3. Lake Michigan is 581 feet higher than the level of the ocean, and Great Salt Lake is 4200 feet above the level of the ocean. How much higher is Great Salt Lake than Lake Michigan?

4. The Caspian Sea is 84 feet lower than the level of the ocean, and the Dead Sea 1290 feet lower than the level of the ocean. How much lower is the Dead Sea than the Caspian Sea?

5. How much higher is Great Salt Lake than the Dead Sea?

6. Pikes Peak is 14,147 feet high. If we allow $\frac{1}{4}$ more for the extra distance due to the slant of its side, how long would it take to walk to its summit at the rate of $1\frac{1}{4}$ miles an hour?

7. The Philippines comprise about 1200 islands with a total area of 114,361 square miles. About what is the average size of the islands?

8. The Mississippi River, from the source of the Missouri to the Gulf of Mexico, is about 4500 miles long. How many days would it take a chip to float from its source to its mouth at an average rate of 3 miles an hour?

9. The Hawaiian Islands contain 6449 square miles, and in 1900 their population was 154,000. What was the population per square mile?

10. The island of Cuba contains 45,884 square miles and 1,572,797 people. How does the population of Cuba per square mile compare with that of the Hawaiian Islands?

Triangles

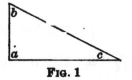 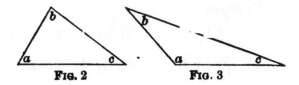

| Fig. 1 | Fig. 2 | Fig. 3 |

A **triangle** is a plane figure having three angles and three sides. The sum of the three angles of any triangle is 180°.

1. In the right triangle (Fig. 1) how many degrees are there in the angle *a*?

2. What must be the sum of the angles *b* and *c*?

3. If the angle *b* is an angle of 60°, how large is the angle *c*?

4. Is it possible for either the angle *b* or the angle *c* to equal 90°?

5. If the angle *b* were to be increased so as to be nearly 90°, what effect would this have upon the shape and size of the triangle?

6. In Fig. 2, since *a* is less than a right angle, must the sum of *b* and *c* be greater or less than 90°?

7. If *a* is 65° and *b* is 85°, what is the size of *c*?

8. In Fig. 3 what would be the effect of increasing the angle *a* to nearly 180°?

9. If *a* is 130°, what is the sum of *b* and *c*?

10. If *a* is 130° and *c* is 30°, how large is *b*?

11. In a right triangle one of the acute angles measures 22°. How large is the other acute angle?

12. In an obtuse-angled triangle the two acute angles measure 18° and 51°. What is the size of the obtuse angle?

Multiplication and Division

Multiply:

1. $\frac{4}{5}$ by 9   **3.** 21 by $\frac{2}{3}$ **5.** $8\frac{2}{7}$ by 7   **7.** $12\frac{1}{2}$ by 8

2. $\frac{7}{15}$ by 8   **4.** 42 by $\frac{4}{7}$ **6.** 28 by $3\frac{1}{14}$ **8.** $10\frac{1}{2}$ by 16

Divide:

9. $\frac{1}{5}$ by $\frac{1}{15}$ **11.** $1\frac{4}{15}$ by 7   **13.** 3 by $1\frac{1}{2}$ **15.** $2\frac{1}{7}$ by $\frac{1}{14}$

10. $3\frac{2}{5}$ by $\frac{1}{5}$ **12.** $\frac{6}{7}$ by 3   **14.** $6\frac{1}{4}$ by $1\frac{1}{4}$ **16.** $8\frac{4}{7}$ by $2\frac{1}{7}$

17. How much will $6\frac{3}{4}$ pounds of sugar cost at 4 cents a pound?

18. If 3 tons of coal cost 15\frac{9}{10}$, what is the price per ton?

19. If 3 tons of coal cost 18\frac{3}{10}$, how much will 8 tons cost?

20. If $2\frac{1}{2}$ yards of cloth cost 25 cents, how much will $7\frac{1}{5}$ yards cost?

21. What is $\frac{1}{5}$ of $10\frac{5}{8}$ bushels?

22. What is $\frac{3}{4}$ of $24\frac{4}{5}$ gallons?

23. How many feet are there in $\frac{1}{4}$ of $16\frac{8}{15}$ feet?

24. If $3\frac{3}{4}$ is $\frac{3}{4}$ of a number, what is the number? What is $4\frac{2}{5}$ times the number?

25. If $8\frac{4}{5}$ is $\frac{4}{5}$ of a number, what is $3\frac{7}{11}$ times the number?

26. If a man earns $7 in $3\frac{1}{2}$ days, how much will he earn in $7\frac{1}{4}$ days?

27. At the rate of 60 cents a bushel, how much will $3\frac{1}{2}$ pecks of potatoes be worth?

28. A man exchanged 2 barrels of apples at $1\frac{1}{2}$ a barrel for cloth at 15 cents a yard. How many yards did he receive?

79

Miscellaneous Problems

1. How many ounces are there in $5\frac{3}{4}$ pounds?

2. How many feet are there in 3 rods?

3. How many yards are there in $6\frac{1}{2}$ rods?

4. What is the cost of $2\frac{1}{4}$ pks. of peanuts at $4\frac{1}{2}$ cents a quart?

5. If milk costs 7 cents a quart, and I take from the milkman 2 quarts and a pint of milk a day, what will be my milk bill for the month of January?

6. How many beds, each containing 20 sq. ft., can be marked off upon a square rod of land, and how many square feet will remain?

7. What part of the money received for goods is profit when goods are sold for $\frac{6}{5}$ of their cost?

8. What part of the cost does the profit equal when goods are sold for $\frac{6}{5}$ of their cost?

9. What part is lost by selling goods for $\frac{5}{8}$ of their cost?

10. If I sell goods so as to gain 25 cents, and by so doing gain $\frac{1}{4}$ as much as the cost, what was the cost?

11. If I lose 20 cents a yard on cloth, and this loss is equal to $\frac{1}{5}$ of the cost, what was the cost?

12. A man leaves $\frac{1}{2}$ of his property to his wife, $\frac{1}{3}$ to his son, and the remainder to his daughter. What part of the property does the daughter receive?

13. If a man leaves $\frac{2}{5}$ of his estate to his wife, $\frac{3}{10}$ to his daughter, and the remainder, which amounts to $3000, to his son, what was the amount of the estate?

14. A can do a piece of work in 4 days, and B in 8 days. What part of it can both together do in a day?

First find what part of the work each alone can do in a day.

Multiplication of Mixed Numbers

1. Multiply $6\frac{1}{2}$ by $4\frac{1}{4}$.

$$\begin{array}{r} 6\frac{1}{2} \\ \underline{4\frac{1}{4}} \\ 24 \\ 2 \\ 1\frac{1}{2} \\ \underline{\frac{1}{8}} \\ 27\frac{5}{8} \end{array}$$

In multiplying numbers which contain fractions, each part of the multiplicand is multiplied by each part of the multiplier, and these products are added. 4 times 6 are 24. 4 times $\frac{1}{2}$ are 2. $\frac{1}{4}$ of 6 is $1\frac{1}{2}$. $\frac{1}{4}$ of $\frac{1}{2}$ is $\frac{1}{8}$.

Multiply:

2.	**3.**	**4.**	**5.**	**6.**
$8\frac{1}{2}$	$7\frac{1}{4}$	$10\frac{2}{3}$	$6\frac{3}{5}$	$8\frac{4}{5}$
$2\frac{1}{2}$	$6\frac{1}{2}$	$3\frac{1}{2}$	$5\frac{1}{3}$	$5\frac{3}{4}$

7.	**8.**	**9.**	**10.**	**11.**
$12\frac{1}{3}$	$18\frac{6}{7}$	$16\frac{4}{5}$	$25\frac{5}{12}$	47
$3\frac{1}{2}$	$14\frac{1}{2}$	15	$24\frac{4}{5}$	$15\frac{2}{3}$

12. If a man walks $3\frac{3}{4}$ miles an hour, how many miles will he walk in $4\frac{1}{2}$ hours?

13. I bought $7\frac{5}{8}$ pecks of potatoes at 24 cents a peck. What was the cost?

14. In a mill where the men work $9\frac{3}{4}$ hours a day, how many hours do they work in 15 days?

15. If a man can build $6\frac{3}{5}$ rods of fence in a day, how many rods can he build in $10\frac{1}{3}$ days?

16. William can ride his bicycle 9 miles an hour. How far can he ride in $11\frac{3}{5}$ hours?

17. At the rate of $\$14\frac{1}{2}$ a week how much can a man earn in $5\frac{2}{7}$ weeks?

18. Mr. Barrows had 64 acres of wheat. It yielded $23\frac{7}{8}$ bushels per acre. How many bushels did he raise?

19. How much will $15\frac{2}{3}$ yards of cloth cost at $12\frac{1}{2}$ cents a yard?

Construction

Review pages 71 and 78.

Make the straight lines with a ruler. Make angles and measure angles with a protractor. See page 255.

1. Make a right angle.

2. Make an acute angle.

3. Make an obtuse angle.

4. Make an angle of 30°.

5. Make an angle of 100°.

6. Make two straight lines crossing each other. Measure the angles. Are any of the angles equal?

7. Make a straight line 3 inches long. Make an angle of 40° at one end. Make an angle of 80° on the same side at the other end. Prolong the lines until they meet. Measure the third angle. Find the number of degrees in all the angles together.

8. Make a straight line $2\frac{1}{2}$ inches long. Make a right angle at one end of the line. Make an angle of 45° at the other end on the same side. Find the sum of all the angles.

9. Make a straight line 2 inches long. Make an angle of 90° at each end of the line and on the same side. Make the sides of the angles 2 inches long. Draw a line connecting the extremities of these lines. Measure the two angles last formed.

10. Construct a square $1\frac{1}{2}$ inches long.

11. Make a horizontal straight line 3 inches long. At each end and on the same side of the line make a right angle. Make the perpendicular lines $2\frac{1}{4}$ inches long and connect their extremities.

12. Make an oblong $2\frac{3}{4}$ inches long and $1\frac{3}{4}$ inches wide.

Review Problems

1. If I buy oranges at the rate of 3 for 2 cents and sell them at the rate of 3 for 4 cents, how much shall I make on a dozen?

2. If I buy oranges at the rate of 5 for 3 cents and sell them at the rate of 3 for 5 cents, how much shall I make on 5 dozen?

3. If I buy 60 yards of cloth for $6, and sell it at $12\frac{1}{2}$¢ a yard, what is the whole profit?

4. If it costs 6 cents an evening for gas for 3 burners, how much will it cost for 5 burners for the month of November?

5. If the average daily circulation of a newspaper is 56,240 copies, how much would the daily sales amount to at 2 cents a copy?

6. The average daily circulation of a newspaper published week days is 85,600. The average wholesale price is $1\frac{1}{4}$ cents per copy. What is the total amount of receipts for a year?

7. If a board is 2 feet wide, how long must it be to contain 30 feet of lumber?

8. If a board is 18 inches wide, how long must it be to contain 12 feet?

9. If a board is 20 feet long, how wide must it be to contain 10 feet?

10. If the 4th of July is Tuesday, what day of the week is the first day of August?

11. What day of the month is 5 weeks later than the 4th of July?

12. The first day of January 1900 was Monday. What day of the week was Washington's Birthday of the same year?

Time between Dates

1. Find the time from May 15, 1895, to Sept. 2, 1899.

FIRST METHOD. From May 1895 to May 1899 is 4 years. From May 15, 1899, to Aug. 15, 1899, is 3 months. From Aug. 15 to Sept. 2 is 18 days. The time is 4 yr. 3 mo. 18 da.

SECOND METHOD.

1899 yr.	9 mo.	2 da.
1885 yr.	5 mo.	15 da.
4 yr.	3 mo.	17 da.

September is the 9th month in the year and May the 5th month. Writing the dates as compound quantities and subtracting, reckoning 30 days for a month, we have 4 yr. 3 mo. 17 da. The difference in the two answers is due to the fact that the month of August has 31 days instead of 30 days.

It is customary to use the first method when it is necessary to know the exact number of days. The second method is frequently used for ordinary purposes.

Find the time by the first method:

2. From June 10, 1856, to Sept. 23, 1875.

3. From July 20, 1865, to Oct. 10, 1887.

4. From Jan. 25, 1871, to Mar. 20, 1892.

5. From Nov. 13, 1858, to Jan. 30, 1881.

6. From Dec. 31, 1869, to Apr. 1, 1900.

7. From Aug. 18, 1891, to Nov. 5, 1898.

8. From Apr. 13, 1892, to May 1, 1901.

Find the time by the second method:

9. From Oct. 15, 1872, to Dec. 25, 1891.

10. From June 22, 1843, to Aug. 6, 1868.

11. From Feb. 12, 1873, to Jan. 18, 1898.

12. From Aug. 8, 1842, to June 5, 1897.

13. From Oct. 18, 1881, to July 12, 1901.

14. From Jan. 9, 1856, to May 1, 1889.

15. From Feb. 27, 1892, to Mar. 3, 1901.

Area of Triangles

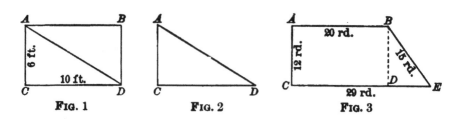

FIG. 1   FIG. 2   FIG. 3

In Fig. 1 observe that the diagonal *AD* divides the rectangle into two equal triangles.

1. What is the area of the rectangle *ABDC* ?

2. What is the area of the triangle *ADC*?

The **altitude** of a triangle is the perpendicular distance between the base line and the angle opposite.

3. What is the area of a triangle whose base is 15 feet and whose altitude is 12 feet ?

4. What is the area of a lot of land in the form of a right triangle whose two sides forming the right angle are 40 rods and 36 rods ?

5. In Fig. 3 what is the distance from *D* to *E*? From *B* to *D* ?

6. What is the area of the rectangular part *ABDC*?

7. What is the area of the triangular part *BED*?

8. If a rectangular lot of land 80 rods long and 30 rods wide is divided by a diagonal line, how many acres are there in one part ?

9. A farm in the form of Fig. 3 is 60 rods wide, 80 rods long on one side, and 100 rods long on the other. How many acres are there in the triangular projection corresponding to *BED*?

10. How many acres are there in the whole farm ?

11. If the farm were divided by a line from *B* to *C*, how many acres would there be in the part *BEC*?

Sound

1. On a clear, mild day sound travels in the air at the rate of about 1120 feet per second. How long does it take sound to travel a mile?

2. When the temperature is at the freezing point, sound travels in the air at the rate of about 1090 feet per second. About how long does it take sound to travel 5 miles on a cold day?

3. If I see the steam from the whistle of a steamboat, and 3 seconds later hear the sound of the whistle, about how far away is the steamboat?

4. In watching a woodsman chopping upon a distant hill, I notice that he strikes exactly one blow per second. After he has struck the last blow, I still hear the report of two blows. About how far distant is he?

5. If 5 seconds elapse between the flash and the sound of a discharge of lightning, how far distant is the discharge?

6. If the echo of the voice reflected from a cliff is heard $1\frac{1}{2}$ seconds after shouting, how far distant is the cliff?

7. If men were stationed 1120 feet apart, how many men would be necessary to cover the distance of 2 miles, 640 feet?

8. If signal guns were stationed 5 miles, 480 feet apart, and it should take a gunner 2 seconds to discharge a gun after hearing the sound of the preceding gun, how long would it take to send a signal in this way from Boston to Chicago, a distance of 1038 miles?

9. In water sound travels at the rate of 4708 feet per second. How long does it take sound to travel 10 miles in water?

Measurements

1. A picture, measured outside the frame, is 30 inches long and 24 inches wide. The frame is 4 inches wide. How far is it around the picture inside the frame?

2. How many square inches are there in the frame?

3. How many square inches in the whole pictûre?

4. A garden 4 rods long and 2 rods wide is surrounded by a concrete walk 3 feet wide. How many feet is it around the garden inside the walk?

5. How many yards is it around the garden outside the walk?

6. A farm is 160 rods long and 80 rods wide. What would be the width of a strip taken from one end, wide enough to make 2 acres?

7. How many acres would there be in a strip 5 rods wide taken from one side?

8. How many square rods would there be in a strip 10 rods wide taken from one side and one end?.

9. A room 18 feet long and 12 feet wide has a carpet with a border. The border is 30 inches wide. What is the distance around the carpet inside the border?

10. Find the number of square yards in the carpet inside the border.

Construction

Review pages 82 and 85. See page 255.

1. Construct a right triangle with one angle of 30°. How large should the remaining angle be? Measure it.

2. Construct a triangle with one angle of 48° and another of 72°. How large should the remaining angle be? Measure it.

3. Construct a triangle containing two acute angles. Measure the angles and find their sum.

4. Construct a triangle containing an obtuse angle. Find the sum of two of the angles and determine how large the third angle should be. Measure it.

5. If it is possible, construct a triangle containing two angles of 90° each.

6. Construct a square. Draw the diagonal. Measure the angles formed. Which angles are equal to each other?

7. Construct a rectangle. Draw the diagonal. Measure the angles. Which angles are equal to each other?

8. Make an angle of 90°. Make the sides 2 inches long. Connect them. Complete the square.

9. Make an angle of 45°. Make one side $2\frac{1}{2}$ inches long. At the end of this line make a right angle and prolong its side. Complete the square.

10. Make an angle of 30°. Make one of the sides 3 inches long. At the end of this line make a right angle and prolong its side. Complete the triangle.

11. Construct a triangle with a side $2\frac{3}{4}$ inches long and an angle of 75°.

12. Construct a triangle with a side $1\frac{3}{8}$ inches long and an angle of 160°.

1. Multiply $12\frac{1}{2}$ by $12\frac{1}{2}$.
2. Multiply 24 by $9\frac{3}{4}$.
3. Multiply $27\frac{5}{6}$ by 30.
4. Multiply $15\frac{4}{9}$ by $18\frac{1}{5}$.
5. Multiply $21\frac{7}{8}$ by 7.
6. Multiply 30 by $6\frac{4}{5}$.

7. $16\frac{1}{2} \times 16\frac{1}{2} = ?$
8. $24\frac{3}{4} \times 12\frac{3}{8} = ?$
9. $34\frac{5}{8} \times 10 = ?$
10. $80 \times 17\frac{3}{10} = ?$
11. $35\frac{7}{8} \times 32\frac{5}{7} = ?$
12. $63\frac{9}{10} \times 40\frac{2}{8} = ?$

13. Divide $25\frac{5}{8}$ by 5.
14. Divide 9 by $4\frac{1}{2}$.
15. Divide 18 by $4\frac{1}{2}$.
16. Divide 7 by $2\frac{1}{3}$.
17. Divide 14 by $2\frac{1}{8}$.
18. Divide 10 by $3\frac{1}{3}$.

19. $\frac{14}{15} \div 7 = ?$
20. $\frac{14}{15} \div \frac{2}{15} = ?$
21. $36\frac{9}{14} \div 3 = ?$
22. $6\frac{9}{10} \div \frac{8}{10} = ?$
23. $2\frac{4}{15} \div \frac{2}{15} = ?$
24. $40 \div \frac{5}{8} = ?$

Find the area of:
25. Rectangle 14 ft. long, 2 yd. wide.
26. Rectangle 7 yd. long, $4\frac{1}{3}$ ft. wide.
27. Rectangle $8\frac{1}{3}$ yd. long, 5 ft. wide.
28. Triangle whose base is 12 ft. and altitude 8 ft.
29. Triangle whose base is $6\frac{1}{2}$ ft. and altitude 10 ft.
30. Triangle whose base is 15 ft. and altitude $6\frac{2}{3}$ ft.

Find the distance around the inside of a picture frame:
31. 6 inches wide and 4 ft. × 2 ft. outside.
32. 4 inches wide and $3\frac{1}{2}$ ft. × $2\frac{1}{2}$ ft. outside.
33. 2 inches wide and $15\frac{1}{3}$ in. × $9\frac{2}{3}$ in. outside.

Find the distance around the outside of a frame:
34. 5 inches wide and 20 in. × 15 in. inside.
35. $4\frac{1}{2}$ inches wide and 3 ft. × $2\frac{1}{2}$ ft. inside.
36. $2\frac{1}{4}$ inches wide and 20 in. × 10 in. inside.

Original Problems

Make problems and solve them:

1. A right angle is divided into two equal parts.

2. A right angle is taken from an angle of 125°.

3. A certain field contains 4 acres and is 32 rods long.

4. A tight board fence is to be built around a yard $2\frac{1}{2}$ rods long and $1\frac{1}{2}$ rods wide.

5. In a right triangle one of the angles measures 25°.

6. Two angles of a triangle measure respectively 23° and 104°.

7. A clothier gained $3 by selling a coat for $\frac{1}{6}$ more than the cost.

8. A melon and an orange are together worth 12 cents.

9. The daily circulation of a certain newspaper is 50,000 copies.

10. The Fourth of July of a certain year was Wednesday.

11. A boy steps 2 feet each time and takes 160 steps a minute.

12. The two shorter sides of a right triangle are 12 feet and 8 feet.

13. The steam from the whistle of a locomotive is seen 4 seconds before the sound is heard.

14. I hear the echo of my voice 2 seconds after shouting.

15. A picture frame is 3 inches wide. The outer dimensions of the frame are 20 in. × 15 in.

16. A garden is surrounded by a walk 4 feet wide.

17. The altitude of a triangle is 4 ft. 6 in. and its base 8 ft.

18. The base of a parallelogram is 12 feet and its area 96 feet.

90

Drill Table

A large number of examples for practice may be obtained from this page by adding the columns as far as any given point; by dividing a column and from the sum of one part taking the sum of the other part; by proceeding both vertically and horizontally, etc.

	1.	2.	3.	4.	5.
a.	4387	1249	6534	2121	8334
b.	8968	2724	7841	5385	7867
c.	9457	6352	6773	3198	5724
d.	3923	8475	7864	2184	1261
e.	7556	3319	7977	9465	4348
f.	4832	6918	6439	6875	5988
g.	5836	1486	5974	2375	3842
h.	6342	4814	2793	8658	7923
i.	8421	9783	6314	4683	5392
j.	3236	7894	5406	4685	9793
k.	3864	6007	4874	8995	8324
l.	6486	7875	2438	6547	9735
m.	4573	3785	8119	9041	2755
n.	8825	5617	4198	1338	3319
o.	6487	2335	6545	8000	5467
p.	4653	9821	8981	2989	6842
q.	5426	3479	8236	9847	3645
r.	3482	7845	3744	6387	1035
s.	3126	5684	8915	3461	1253
t.	5391	1936	7532	2183	7928
u.	7030	4236	4958	3863	2468
v.	7924	8117	7123	2735	5256
w.	1803	3261	4138	3971	6453
x.	7158	7934	2194	7861	9645
y.	2139	4861	2537	7653	2932
z.	4130	1423	4270	4115	9742

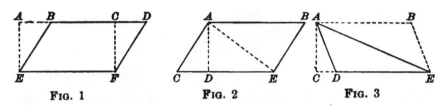

FIG. 1   FIG. 2   FIG. 3

A plane figure with four sides, whose opposite sides are parallel to each other, is called a **parallelogram**.

The figure *BDEF* is a **parallelogram**. The line *EF*, upon which it is supposed to rest, is called the **base**, and the height, measured by the line *FC*, is called the **altitude**.

In Fig. 1 observe that by changing the position of the parts of the parallelogram it may be changed to a rectangle having the same base and altitude.

1. What is the area of a parallelogram whose base is 12 feet and altitude $8\frac{1}{4}$ feet ?

2. How many square rods are there in a piece of land in the form of a parallelogram, if two opposite parallel sides are each 36 rods long, and the distance between them is $12\frac{2}{3}$ rods ?

Observe by Fig. 2 that any parallelogram may be divided into two equal triangles, and by Fig. 3 that any triangle may be regarded as half of a parallelogram having the same base and altitude.

3. What is the area of a triangle whose base is 9 yards and altitude $3\frac{2}{3}$ yards ?

4. A lot of land is in the form of the triangle *AEC* (Fig. 2). The side *CE* is 100 rods ; the distance from *D* to *E* is 70 rods ; and the distance *AD* 40 rods. How many square rods are there in the part *ADE* ?

5. How many square rods in the part *ADC* ?

6. How many square feet are there in a flower bed in the form of a right triangle, if the two sides forming the right angle are 10 ft. 8 in. and 9 ft. 6 in. ?

92

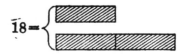

1. If I should divide 18 oranges between John and James, giving John twice as many as James, how many would each receive?

Every time James receives 1, John receives 2, and together they receive 3; John therefore receives ⅓ and James ⅓ of the oranges.

2. Divide the number 16 into two parts so that one shall be three times as large as the other.

1 + 3 = 4. ¼ of 16 = ? ¾ of 16 = ?

3. Divide the number 25 into two parts so that one shall be 4 times as large as the other.

4. If I divide 15 sheets of paper between a boy and a girl, giving the boy 2 sheets and the girl 3 sheets, repeating this until all has been divided, how many sheets will each receive?

5. How should I divide 24 foreign stamps between 2 boys so as to give one boy 5 stamps as many times as I should give the other boy 3 stamps?

6. If $36 were divided between 2 men in the proportion of 5 and 4, how many would each receive?

7. Divide the number 44 into 2 parts in the proportion of 6 and 5.

8. Two boys 300 miles apart start on their bicycles to ride towards each other. One boy rides 8 miles an hour, on the average, and the other 7 miles an hour. How many hours will it be before they will meet? How many miles will each have ridden when they meet?

9. Divide the number 10 into two parts in the proportion of 20 and 30.

The proportion of 20 and 30 is the same as the proportion of 2 and 3.

Accounts

EDWARD KNOWLES,

BOUGHT OF HOME FURNISHING CO.

Apr.	20	1 Dining Table,	$12	00		
"	20	8 Chairs @ $1.40,	11	20		
"	20	32 yd. Carpeting @ 80¢,	25	60		
"	20	1 Desk,	15	50		
"	20	1 Lamp,	3	25	$67	55

Received payment,

HOME FURNISHING CO.

PER GATES.

Make bills for the following accounts and receipt them.

Make the lines as in the above bill.

1. March 5, 1900, William Sherman bought of Andrew Swan & Co. 12 yd. flannel @ 32¢; $9\frac{1}{2}$ yd. dress goods @ 40¢; 6 spools thread @ 5¢; 4 pr. socks @ 15¢.

2. October 1, 1900, Chas. H. Peters sold to W. F. Moore 20 bu. corn @ 72¢; 50 bu. oats @ 38¢; 12 bbl. flour @ $4.75.

3. June 5, 1900, A. T. Lyman & Co. bought of Smith & Fales 10 lb. coffee @ 37¢; 40 lb. butter @ 22¢; 1 doz. oranges, 30¢; 25 lb. sugar @ 5¢.

4. July 30, 1900, P. H. Sayles & Co sold E. H. Johnson 1 bbl. pork, $18.00; 125 lb. beef @ 9¢; 60 lb. lard @ $7\frac{1}{2}$¢.

5. Feb. 1, 1901, J. B. Williams bought of A. H. Brown & Co. $527\frac{1}{2}$ lb. butter @ 24¢; 312 lb. cheese @ $12\frac{1}{4}$¢; 98 doz. eggs @ $22\frac{1}{2}$¢.

6. March 5, 1901, Wilbur & Jackson sold Mrs. E. P. Stillman $68\frac{1}{2}$ yd. cotton cloth @ $9\frac{1}{2}$¢; 1 pair gloves, 85¢; $13\frac{3}{4}$ yd. gingham @ $8\frac{1}{3}$¢; 4 pair hose @ 35¢.

1. How much is one ninth and one eighteenth? Four ninths and five eighteenths? Five sixths less one eighteenth?

2. What is the sum of one sixth and one ninth? Five sixths and two ninths? One third, one sixth, and one ninth?

3. How much is one third, one ninth, and one eighteenth? One tenth and one twentieth? One fifth, one tenth, and one twentieth?

Add:

4.	**5.**	**6.**	**7.**	**8.**
$4\frac{5}{9}$	$12\frac{5}{6}$	$25\frac{1}{5}$	$42\frac{2}{8}$	$58\frac{19}{20}$
$6\frac{1}{3}$	$8\frac{2}{3}$	$34\frac{1}{2}$	$63\frac{1}{2}$	$65\frac{3}{10}$
$7\frac{5}{18}$	$10\frac{17}{18}$	$17\frac{7}{10}$	$29\frac{7}{20}$	$48\frac{4}{5}$

Subtract:

9.	**10.**	**11.**	**12.**	**13.**
$12\frac{7}{18}$	$20\frac{8}{9}$	$16\frac{1}{3}$	$22\frac{2}{3}$	$32\frac{1}{6}$
$8\frac{2}{9}$	$9\frac{5}{18}$	$11\frac{8}{9}$	$15\frac{5}{18}$	$16\frac{2}{9}$

14.	**15.**	**16.**	**17.**	**18.**
$9\frac{1}{5}$	$10\frac{1}{2}$	$25\frac{7}{20}$	$45\frac{1}{2}$ ·	$86\frac{19}{20}$
$6\frac{3}{20}$	$5\frac{9}{20}$	$10\frac{4}{5}$	$23\frac{9}{10}$	$43\frac{3}{4}$

Multiply:

19. $9\frac{1}{2}$ by $\frac{1}{4}$ **22.** 10 by $7\frac{7}{10}$ **25.** $8\frac{1}{2}$ by $4\frac{1}{4}$ **28.** $20\frac{5}{6}$ by $15\frac{1}{4}$

20. $6\frac{5}{6}$ by 5 **23.** 25 by $10\frac{1}{5}$ **26.** $12\frac{1}{3}$ by $9\frac{1}{2}$ **29.** $40\frac{1}{2}$ by $16\frac{7}{10}$

21. $8\frac{2}{9}$ by 9 **24.** 40 by $8\frac{1}{20}$ **27.** $15\frac{1}{6}$ by $12\frac{1}{3}$ **30.** $23\frac{1}{3}$ by $18\frac{1}{6}$

Divide:

31. $12\frac{8}{9}$ by 4 **33.** $\frac{2}{3}$ by $\frac{1}{6}$ **35.** $3\frac{2}{3}$ by $\frac{1}{3}$ **37.** 3 by $1\frac{1}{2}$

32. $20\frac{5}{18}$ by 5 **34.** $\frac{5}{6}$ by $\frac{5}{18}$ **36.** $8\frac{1}{5}$ by $\frac{1}{5}$ **38.** 10 by $3\frac{1}{3}$

Construction

Review pages 78, 85, and 92.

1. Make a horizontal line 2½ inches long. At one end make an angle of 75° and make the second side of the angle 1½ inches long. At the extremity of the last line and on the same side as the first angle make an angle of 105°. Prolong the line to 2½ inches. Draw a line completing the figure. Measure the angles last formed. Are any of the four angles equal to each other? Are there two angles whose sum is 180°?

2. Construct a parallelogram one of whose sides is 3 inches long and one of whose angles measures 68°. Which angles are equal to each other? Which two angles would make 180° if combined? Are any of the sides equal to each other? Make a right angle upon one side with dotted lines and measure the altitude of the parallelogram. What is the area of the parallelogram?

3. Construct a parallelogram having an angle of 45° and a side 2¾ inches long. Draw the diagonal of the parallelogram. Measure the angles on both sides of the diagonal. Which are equal to each other? Make a perpendicular dotted line representing the altitude of one of the two triangles formed. Find the area of the triangle.

4. Make a triangle having a side 4 inches long, an angle of 65°, and another angle of 40°. How large must the third angle be? Measure the altitude and reckon the area of the triangle.

5. Make a triangle having a side 3½ inches long, an angle of 100°, and another angle of 30°. Regarding this triangle as one half of a parallelogram, complete the parallelogram. Measure the altitude and find the area of the parallelogram.

Cubes

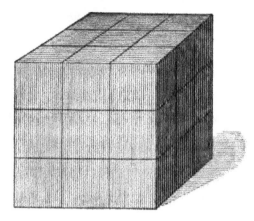

A block one inch long, one inch wide, and one inch high is called a **cubic inch.**

1. How many sides or faces are there on a cubical block?

2. How many different straight edges are there upon a cube?

3. If 3 cubes are set in a row, and 2 more rows are placed beside these, how many cubes will there then be together?

4. How many inch cubes can be placed upon a square surface 5 in. long and 5 in. wide?

5. How many similar layers would have to be added to make a cube 5 in. high?

6. How many inch cubes can be placed upon a square surface 12 in. long and 12 in. wide?

7. How many similar layers would have to be placed one upon another to make a cube 12 in. high?

8. How many cubic inches are there in a cubic foot?

9. How many cubic feet can be placed upon a surface a yard square?

10. How many layers must be built up to make a cube a yard high?

Problems

1. Into how many 2-in. cubes can a 4-in. cube be divided?

2. How many 2-in. cubes are contained in a 6-in. cube?

3. How many 3-in. cubical blocks can be packed in a box a foot long, 6 inches wide, and 9 inches deep?

4. A mass of candy in the form of a cube 6 inches long is cut into cubes 3 inches long. How many will it make?

5. How many times as large is a cube 3 feet long as a cube 3 inches long?

6. How many times as large is a cube 3 yards long as a cube 3 feet long?

7. A cubic foot of pure water weighs 1000 ounces. How many pounds will 3 cubic feet of water weigh?

8. How many cubic feet are there in a tank in the form of a cube which is 4 feet long?

9. How many pounds of water would this tank contain?

10. Iron weighs about 7 times as much as water. What would be the weight of 4 cubic feet of iron?

11. Silver weighs about $10\frac{1}{2}$ times as much as water. Find about how much a cubic inch of silver weighs.

12. What would be the weight of a cubical piece of silver 3 inches long?

13. Gold weighs about $19\frac{1}{4}$ times as much as water. About what would be the weight of a cubic foot of gold?

14. Find whether a cubic inch of gold would weigh more or less than a pound.

15. Mercury weighs about $13\frac{1}{2}$ times as much as water. Find the weight of $\frac{1}{2}$ of a cubic foot of mercury.

Fractions — Drill Work

Add:

1. $12\frac{2}{3} + 17\frac{1}{2} + 38\frac{5}{6}$

2. $27\frac{1}{5} + 39\frac{3}{10} + 46\frac{1}{2}$

3. $18\frac{3}{4} + 58\frac{2}{3} + 27\frac{5}{6}$

4. $46\frac{3}{7} + 65\frac{1}{2} + 73\frac{5}{14}$

5. $82\frac{1}{3} + 19\frac{4}{5} + 38\frac{14}{15}$

6. $68\frac{1}{2} + 43\frac{5}{9} + 21\frac{1}{6}$

7. $285\frac{3}{8} + 367\frac{7}{12}$

8. $921\frac{1}{2} + 1505\frac{5}{9}$

9. $3425\frac{5}{8} + 2401\frac{5}{16}$

10. $520\frac{11}{12} + 7305\frac{5}{6}$

11. $4994\frac{1}{2} + 3898\frac{8}{9}$

12. $7481\frac{1}{5} + 6422\frac{2}{3}$

Subtract:

13. $75\frac{2}{3} - 60\frac{1}{2}$

14. $83\frac{1}{3} - 72\frac{3}{4}$

15. $95\frac{7}{8} - 21\frac{3}{16}$

16. $72\frac{1}{8} - 33\frac{15}{16}$

17. $80\frac{3}{7} - 30\frac{3}{14}$

18. $71\frac{3}{9} - 22\frac{3}{18}$

19. $321\frac{1}{2} - 84\frac{5}{14}$

20. $248\frac{1}{4} - 110\frac{2}{3}$

21. $562\frac{2}{9} - 493\frac{1}{18}$

22. $480\frac{6}{7} - 280\frac{13}{14}$

23. $756\frac{3}{8} - 456\frac{1}{16}$

24. $989\frac{1}{18} - 279\frac{8}{9}$

Multiply:

25. $375 \times 35\frac{4}{5}$

26. $602 \times 28\frac{6}{7}$

27. $528 \times 39\frac{11}{12}$

28. $272\frac{1}{4} \times 84$

29. $156\frac{13}{14} \times 70$

30. $871\frac{17}{18} \times 90$

31. $25\frac{1}{2} \times 36\frac{4}{5}$

32. $56\frac{2}{3} \times 48\frac{3}{4}$

33. $77\frac{1}{2} \times 62\frac{4}{7}$

34. $64\frac{1}{5} \times 55\frac{5}{8}$

35. $36\frac{2}{7} \times 42\frac{7}{12}$

36. $54\frac{1}{11} \times 44\frac{11}{18}$

Divide:

37. $2455\frac{5}{8} \div 5$

38. $3608\frac{8}{9} \div 8$

39. $570\frac{13}{18} \div 3$

40. $637\frac{14}{16} \div 7$

41. $485\frac{15}{16} \div 5$

42. $527\frac{17}{18} \div 17$

43. $567\frac{6}{7} \div 3$

44. $568\frac{13}{18} \div 4$

45. $360\frac{10}{15} \div 5$

46. $252\frac{7}{16} \div 7$

47. $416\frac{16}{17} \div 8$

48. $918\frac{18}{19} \div 9$

Decimal Fractions

Tenths are so frequently used that a more convenient method of writing them has been adopted. A figure placed at the right of a dot is regarded as indicating tenths. A fraction written in this way is called a **decimal fraction**. The dot is called a **decimal point**. $\frac{7}{10}$ in the decimal form is .7; $5\frac{3}{10}$ is 5.3.

1. Add 8.4, 5, and 3.9.

8.4	Place all the tenths in the same column.
5	9 tenths and 4 tenths are 13 tenths, or 1 whole and 3 tenths.
3.9	
17.3	One whole with 16 wholes makes 17 wholes.

2. From 16.3 take 8.5.

16.3	5 tenths from 13 tenths leaves 8 tenths. 8 wholes from 15
8.5	wholes leaves 7 wholes.
7.8	

Add:

3.	4.	5.	6.	7.
4.8	6.9	25.1	64.3	36
42	4.1	18.4	82.9	8.3
3.9	21	45.3	9.7	41.7
.7	7.3	16	12.1	15.4

Subtract:

8.	9.	10.	11.	12.
9.9	13	24.7	36	45
7.2	8.5	12.3	24.9	21.3

13. Add eighteen and seven tenths; one hundred twenty-five; three and five tenths; eight tenths; thirty-five and one tenth.

14. From thirty-eight and four tenths take seventeen.

15. From one hundred take fifty and three tenths.

Decimal Fractions

1. Multiply 3.7 by 8.

$$\begin{array}{r} 3.7 \\ 8 \\ \hline 29.6 \end{array}$$

8 times 7 tenths are 56 tenths, or 5 and 6 tenths. 8 times 3 are 24. 24 and 5 are 29.

2. Divide 25.6 by 8.

$$\begin{array}{r} 8)\overline{25.6} \\ 3.2 \end{array}$$

25 and 6 tenths is the same as 256 tenths. 256 tenths divided by 8 is 32 tenths, or 3 and 2 tenths.

Multiply :

3. .8 by 4.

4. 5.3 by 7.

5. 9.8 by 7.

6. 8.7 by 6.

7. 17.4 by $4\frac{1}{2}$.

8. 12.6 by $7\frac{2}{3}$.

9. 42.5 by $5\frac{4}{5}$.

10. 72.8 by $9\frac{1}{8}$.

11. 36.5 by 15.

12. 28.7 by 47.

13. 82.1 by 29.

14. 65.9 by 85.

Divide :

15. 18.6 by 3.

16. 34.5 by 5.

17. 19.2 by 8.

18. 31.5 by 7.

19. .9 by 3.

20. .5 by $2\frac{1}{2}$.

21. 2.5 by $12\frac{1}{2}$.

22. 1.7 by $4\frac{1}{4}$.

23. 14.4 by 12.

24. 25.5 by 15.

25. 30.6 by 17.

26. 62.5 by 25.

27. Find the combined length of 24 steel rods each of which is 9.8 inches long.

28. Find $\frac{1}{8}$ of the distance between two places which are 95.2 miles apart.

29. During a certain storm 1.3 inches of rain fell. If there were a storm like this every week in the year, what would be the total amount for the year?

30. If the annual rainfall of a place is 32.4 inches, how much is the average for each month?

101

Rectangular Solids

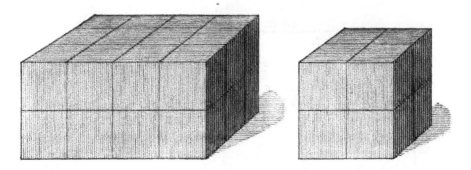

1. How many cubical blocks an inch long would be required to cover a space 4 inches long and 3 inches wide?

2. If another layer were placed above these, how many inch cubes would there then be in all?

A solid body whose corners all form right angles is called a **rectangular solid**.

3. How many cubic inches are there in a rectangular solid which is 6 inches long, 5 inches wide, and 4 inches thick?

4. How many cubic feet in a rectangular solid 5½ feet long, 4 feet wide, and 3½ feet high?

5. How many cubic feet of water would be contained in a tank 6¼ feet long, 4½ feet wide, and 3 feet deep?

6. What would be the weight of the water in this tank?

7. Iron weighs about 7 times as much as water. Find the weight of a bar of iron 6 inches wide, 3 inches thick, and 10 feet long.

8. How many blocks, each 1 foot long, 6 inches wide, and 6 inches thick, can be packed into a box which is 4 feet long, 2 feet wide, and 2 feet deep?

9. Find the number of cubic feet of capacity of a cart 8 feet long, 4.5 feet wide, and 3.3 feet high.

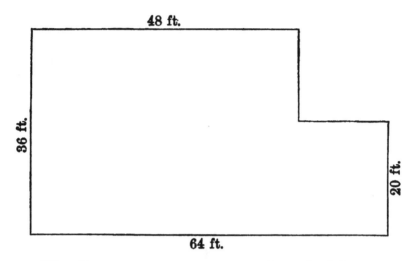

1. The diagram represents ground marked for excavating a cellar. How long is the projecting part?

2. Find the distance entirely around the cellar.

3. How many square yards in the larger part, not including the projection?

4. How many square feet in the whole area?

5. If the cellar is to be excavated to the depth of 9 feet, how many cubic feet of earth will be taken out of the cellar for the projection?

6. How many cubic yards will be taken from the cellar for the part without the projection?

7. What will be the cost of excavating the whole cellar at 37½ cents per cubic yard?

8. How many cubic feet will there be in a wall 18 inches thick built across the west end of the cellar?

9. How many cubic yards will there be in a wall 18 inches thick built entirely around the cellar?

Take the outside measurement for the length of the wall.

10. How much will it cost to build the whole wall at $1.35 a cubic yard?

1. Find the sum of $5\frac{2}{4}$, $6\frac{1}{2}$, and $3\frac{5}{12}$.

2. The sum of 2 numbers is $7\frac{3}{4}$. One of the numbers is $4\frac{1}{4}$. What is the other number?

3. How much would $6\frac{1}{2}$ quarts of milk cost at $4\frac{1}{2}$ cents a quart?

4. At $2\frac{1}{2}$ cents each, how many oranges can be bought for 13 cents?

5. What part of 3 is 2?

6. What part of 9 is 5?

7. What part of $\frac{3}{4}$ is $\frac{1}{4}$?

8. What part of $\frac{3}{4}$ is $\frac{1}{2}$?

9. What part of $\frac{1}{2}$ is $\frac{3}{8}$?

10. 9 is $\frac{3}{4}$ of what number?

11. $\frac{2}{3}$ is $\frac{1}{2}$ of what number?

12. $\frac{6}{7}$ is $\frac{3}{4}$ of what number?

$\frac{1}{4}$ of the number is $\frac{1}{3}$ of $\frac{6}{7}$ or $\frac{2}{7}$; $\frac{4}{4}$ or the whole number is $4 \times \frac{2}{7}$ or $\frac{8}{7}$ or $1\frac{1}{7}$.

13. How many thirds are there in one third more than a whole?

14. How many fifths are there in one fifth less than a whole?

15. 8 is $\frac{1}{3}$ more than what number?

16. 6 is $\frac{1}{4}$ less than what number?

17. If by selling an article for one fifth less than the cost, I lose 5 cents, what was the cost?

5 is $\frac{1}{5}$ of what number?

18. If I sell a pad for one fifth more than the cost, and receive 12 cents for it, what was the cost?

104

1. $\frac{1}{2} + \frac{5}{9} + \frac{1}{18} = ?$

2. $\frac{5}{6} + \frac{1}{3} + \frac{1}{9} = ?$

3. $\frac{3}{4} + \frac{7}{16} + \frac{1}{2} = ?$

4. $\frac{1}{5} + \frac{3}{2} + \frac{9}{10} = ?$

5. $\frac{2}{3} + \frac{2}{5} + \frac{2}{15} = ?$

6. $\frac{2}{9} + \frac{5}{6} + \frac{13}{18} = ?$

7. $\frac{17}{18} - \frac{1}{2} = ?$

8. $\frac{5}{9} - \frac{1}{6} = ?$

9. $\frac{15}{16} - \frac{3}{4} = ?$

10. $3\frac{1}{3} - 1\frac{3}{4} = ?$

11. $7\frac{4}{9} - 3\frac{1}{2} = ?$

12. $10\frac{2}{3} - 4\frac{1}{5} = ?$

13. $\frac{8}{9} \times 10 = ?$

14. $\frac{5}{6} \times 12 = ?$

15. $\frac{3}{5} \times 20 = ?$

16. $12\frac{1}{2} \times 12\frac{1}{2} = ?$

17. $16\frac{2}{3} \times 6 = ?$

18. $8\frac{1}{3} \times 12 = ?$

19. $\frac{7}{18} \div 7 = ?$

20. $\frac{12}{15} \div \frac{2}{5} = ?$

21. $10 \div \frac{5}{8} = ?$

22. $17 \div 4\frac{1}{4} = ?$

23. $10 \div 2\frac{1}{2} = ?$

24. $30 \div 2\frac{1}{2} = ?$

25. Find the volume of an 8-inch cube.

26. Find the volume of a 12-inch cube.

27. Find the volume of a 15-inch cube.

28. Find the volume of a cube $2\frac{1}{2}$ ft. long.

29. Find the volume of a cube $4\frac{1}{2}$ ft. long.

30. Find the volume of a cube 3 ft. 6 in. long.

Find the volume of a rectangular solid:

31. 10 inches long, 8 inches wide, and 5 inches thick.

32. $8\frac{1}{2}$ inches long, $6\frac{1}{2}$ inches wide, and 4 inches thick.

33. 16 inches long, $9\frac{1}{4}$ inches wide, and $7\frac{1}{2}$ inches thick.

34. 5 yards long, 5 feet wide, and 5 feet thick.

35. $4\frac{1}{3}$ yards long, $4\frac{1}{2}$ feet wide, and 2 feet thick.

36. 6 yards long, 6 feet wide, and 6 inches thick.

Original Problems

Make problems and solve them:

1. Divide 15 apples between two boys.

2. Divide the number 24 into two parts.

3. Two men 45 miles apart travel toward each other, one 5 miles an hour and the other 4 miles an hour.

4. Two men engage in trade. One puts $1000 into the business and the other $500.

5. I have 6 apples and give $\frac{2}{3}$ of an apple to each child.

6. Cloth $\frac{3}{4}$ of a yard wide costs 18 cents a yard.

7. I spent $\frac{5}{6}$ of my money and had $12 left.

8. 45 is $\frac{2}{5}$ of a certain number.

9. A 3-inch cube is divided into half-inch cubes.

10. Iron weighs about 7 times as much as water.

11. A box 10 inches long, 5 inches wide, and 3 inches deep is filled with boxes 5 inches long, $2\frac{1}{2}$ inches wide, and $1\frac{1}{2}$ inches deep.

12. A cistern which is full of water is 10 feet long, 6 feet wide, and 4 feet deep.

13. A coal wagon is 8 feet long, 5 feet wide, and $3\frac{1}{2}$ feet high.

14. An excavation is to be made for the cellar of a house 40 ft. long, 25 ft. wide, and 7 ft. deep.

15. It will cost 30 cents a cubic yard to excavate the cellar.

16. The bottom of the cellar is to be covered with concrete.

17. The cellar wall is to be 2 feet thick.

18. The distance around a square field is 20 rods.

19. An oblong is 12 feet long, and the distance around it is 40 feet.

Decimal Fractions

A figure indicating **hundredths** when written in the decimal form occupies the second place to the right of the decimal point.

The fraction one hundredth may be written .01. Seven hundredths may be written .07. Two tenths and three hundredths or twenty-three hundredths would be written .23.

1. Write in the decimal form: nine hundredths; forty-six hundredths; eight, and four hundredths; thirty-five, and three hundredths; eighty-seven, and forty-five hundredths.

2. Add: ten, and fourteen hundredths; eight, and seven hundredths; four, and seven tenths; eight hundredths; twenty-five; three, and three tenths; nine tenths.

See p. 100. Be careful to write all the tenths in the same column and all the hundredths in the same column.

Add:

3.	4.	5.	6.
9.03	.16	13	85.08
23.7	9.5	22.04	20.2
.42	42.07	.8	9.7
5.8	8.4	15.6	93

Subtract:

7.	8.	9.	10.
8.24	9.08	17.4	35
4.12	3.4	8.25	21.13

11.	12.	13.	14.
10.19	35.6	42	100
.28	1.27	.15	5.07

Decimal Fractions

1. Multiply 2.47 by 3.

$$\begin{array}{r} 2.47 \\ 3 \\ \hline 7.41 \end{array}$$

2.47 is the same as 247 hundredths. 3 times 247 hundredths are 741 hundredths, or 7 and 41 hundredths.

2. Divide 9.35 by 5.

$$\begin{array}{r} 5)\overline{9.35} \\ \hline 1.87 \end{array}$$

9.35 is the same as 935 hundredths. 935 hundredths divided by 5 is 187 hundredths, or 1 and 87 hundredths.

Multiply:

3. .25 by 8.   **7.** 30.9 by 8.   **11.** .35 by 15.

4. 8.06 by 9.   **8.** 33.9 by 8.   **12.** .08 by 36.

5. 41.93 by 5.   **9.** .98 by 7.   **13.** 7.58 by 64.

6. 37.55 by 4.   **10.** 9.8 by 7.   **14.** 8.79 by 89.

Divide:

15. .24 by 4.   **19.** .96 by 8.   **23.** 288.24 by 12.

16. 6.35 by 5.   **20.** 9.6 by 8.   **24.** 2882.4 by 12.

17. 12.48 by 8.   **21.** 96 by 8.   **25.** 625.75 by 25.

18. 26.46 by 6.   **22.** 920 by 10.   **26.** 6257.5 by 25.

27. A piece of a surveyor's chain which contains 8 links is 63.36 inches long. How long is one of the links?

28. Find the length of a part of the chain which contains 27 links.

29. An entire chain contains 100 links. How long is it?

30. How many square inches are there in a rectangle that is 19.45 inches long and 13 inches wide?

Review Problems

1. How much shall I make if I buy $7\frac{1}{2}$ gallons of milk at 20 cents a gallon and sell it at 7 cents a quart?

2. How much shall I make by buying $2\frac{3}{4}$ bushels of apples at \$1 a bushel and selling them at 4 cents a quart?

3. A teamster, having 4 horses, gives each horse 4 quarts of oats 3 times a day. How long will 10 bushels last?

4. If a basket of coal weighs 80 pounds, how many baskets will there be in a ton?

5. If a cubic foot of anthracite coal weighs 96 pounds, how many pounds can be put into a cart 10 feet long, $4\frac{1}{2}$ feet wide, and 3 feet high?

6. If a man should step $2\frac{1}{2}$ feet at each step, and should take 150 steps a minute, how long would it take him to walk a mile?

7. How many feet is it around a rectangular lot which is $12\frac{1}{4}$ rods long and $30\frac{2}{3}$ yards wide?

8. How many fence rails 12 feet long will be required to build a fence 3 rails high on both sides of a piece of road 30 rods long?

9. How many boards each 4 ft. long, 6 in. wide will be required to exactly cover the floor of a room which is 12 ft. long and 7 ft. wide?

10. How many stone tiles 9 in. square will it take to lay a floor 9 ft. long and 6 ft. wide?

11. How many cubes 4 in. long can be packed into a cubical box 1 ft. 4 in. long?

12. How many sticks 4 ft. long, 2 in. wide, and 2 in. thick can be packed into a box 4 ft. long, $2\frac{1}{2}$ ft. wide, and $3\frac{1}{8}$ ft. deep?

In making out a bill of accounts, if a bill for certain items has been previously sent, these items are not usually given again, but the whole amount of the previous bill is written first in the new bill. If anything has been paid on the bill, this is subtracted from the amount.

PROVIDENCE, JUNE 5, 1900.

WM. CHANDLER,

TO HENRY MILLER & CO., DR.

		To acct. rendered,			$38	56
May	12	To 3 bbl. Pork @ $18.25,	$54	75		
"	18	" 256 lb. Ham @ 13¢,	33	28	88	03
		Credit,			$126	59
May	22	By cash,			100	00
		To balance,			$26	59

Make monthly statements for the following transactions:

1. George Fifield has an account with Frederick Day & Co. A bill had previously been sent George Fifield, showing an unpaid balance of $85.50. On Apr. 3, 1900, Frederick Day & Co. sold George Fifield 60 lb. sugar @ 6½¢ and 20 lb. coffee @ 35¢, and on Apr. 12, 30 doz. eggs @ 22¢ and 5 bbl. flour @ $5.50. Apr. 25, George Fifield paid cash $90.

2. William Brown, in account with Edward Davis, owes a balance of $12.42, according to a statement which has been sent him. Dec. 8, 1899, he buys on credit 12 yd. cloth @ 42½¢ and 4 pr. socks @ 22¢. On Dec. 17 he buys 1 suit clothes @ $14.50. Jan. 1, 1900, he pays the amount in full. Make out a proper bill and receipt it.

Ratio

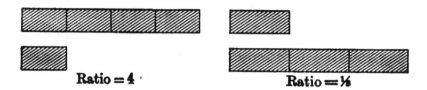

Ratio = 4 · Ratio = ⅙

The **ratio** of one quantity to another is the number of times the first is larger than the second, or the number of times the second is contained in the first. The ratio of 24 to 6 is 4. The ratio of 5 to 15 is ⅓.

1. What is the ratio of 72 to 9?

2. What is the ratio of 8 yards to 6 inches?

3. What is the ratio of a foot square to a 2-inch square?

4. What is the ratio of an 8-inch cube to a 2-inch cube?

5. What is the ratio of a rectangle 12 inches long, 8 inches wide to a rectangle 4 inches long, 2 inches wide?

Find the ratio of the following:

6. 5 gallons, 1 quart to 3 pints.

7. 2 pecks to 2 bushels.

8. 8 rods to 6 feet.

9. An acre to 8 square rods.

10. A square mile to 80 acres.

11. A township 6 miles square to a farm of 320 acres.

12. A lot 400 rods long, 200 rods wide to a farm 200 rods long, 80 rods wide.

13. A cube 3 feet long to a cube 9 inches long.

14. An angle of 175° to an angle of 15°.

Cubic Measure

Find the number of cubic inches in:

1. A cube 10 inches long.
2. A cube 1 foot long.
3. A cube 1 ft. 3 in. long.
4. A cube 1 ft. 6 in. long.
5. A cube 1 ft. 9 in. long.
6. A cube 2 feet long.
7. A cube 2 ft. 4 in. long.
8. A cube 2 ft. 10 in. long.

Find the number of cubic inches in a rectangular solid:

9. 10 in. long, 8 in. wide, 6 in. thick.
10. 1 ft. long, 10 in. wide, 9 in. thick.
11. 2 ft. long, 1 ft. 6 in. wide, 10 in. thick.
12. 1 ft. 8 in. long, 1 ft. 4 in. wide, 1 ft. thick.
13. 2 ft. 4 in. long, 1 ft. 6 in. wide, 1 ft. 2 in. thick.
14. 2 ft. long, 1 ft. wide, $1\frac{1}{2}$ ft. thick.
15. $2\frac{1}{2}$ ft. long, $1\frac{1}{2}$ ft. wide, 1 ft. thick.
16. 3 ft. long, $2\frac{1}{4}$ ft. wide, 1 ft. 4 in. thick.

What is the ratio of:

17. A 2-inch cube to a 4-inch cube?
18. A 2-inch cube to a 6-inch cube?
19. A 4-inch cube to a cube 2 feet long?
20. A 4-inch cube to a cube $4\frac{1}{2}$ feet long?
21. A 3-inch cube to a box 3 ft. × $2\frac{1}{2}$ ft. × 2 ft.?
22. A 2-inch cube to a box 4 ft. × $3\frac{1}{4}$ ft. × $1\frac{1}{2}$ ft.?
23. A block 4 in. × 3 in. × 2 in. to a box 20 in. × 15 in. × 10 in.?
24. A block 2 ft. × 1 ft. × 6 in. to a box 8 ft. × 6 ft. × 4 ft.?

Compound Quantities

1. Find the sum of 4 bu. 3 pk. 2 qt., 3 bu. 2 pk. 3 qt. 1 pt., and 5 bu. 1 pk. 1 qt.

2. From 10 gal. 2 qt. 1 pt. take 4 gal. 3 qt. 1 pt.

3. Multiply 4 yd. 2 ft. 5 in. by 6.

4. Divide 16 lb. 4 oz. by 5.

5. How many pounds are there in 1620 ounces?

6. How many miles in 960 rods?

7. How much will 10 rd. 4 yd. 1 ft. of fence cost at 80 cents a foot?

8. How much will 1750 lb. of hay cost at 90 cents a hundred pounds?

9. What would be a fourth part of 16 sq. rd. and 12 sq. ft.?

10. How many square yards are there in a carpet $16\frac{1}{2}$ ft. long and 12 ft. wide?

11. How many 4-inch cubes will a box hold which is 4 ft. long, 2 ft. wide, and $1\frac{1}{3}$ ft. high?

12. How many cubic inches in a box 4 ft. 2 in. long, 2 ft. 3 in. wide, and 17 in. high?

13. Mt. McKinley, the highest mountain in North America, is 20,464 feet high. How many miles high is it?

14. If one third more than the height is allowed for the inclined distance up the side, how many miles is it up the side of the mountain to its summit?

15. How many feet of 2-inch plank will be required to lay a floor $21\frac{1}{2}$ ft. long and $16\frac{1}{3}$ ft. wide?

16. How many feet of boards will it take to make a box $3\frac{1}{2}$ ft. long, 2 ft. wide, and 2 ft. high, making no allowance for the lapping over at the edges?

Building

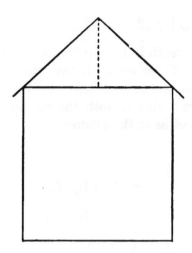

The figure represents the end of a house. The triangular part at the top is called the gable. The height of the gable is measured from the top of the house down to the line of the top of the walls.

1. What is the area of the gable of a house which is 40 ft. wide, if the gable is 12½ ft. high?

2. A house is 60 ft. long and 42 ft. wide. The posts are 20 ft. high and the gables 10 ft. high. How many feet of boards will be required to board the two sides?

3. How many feet will be required to board the two ends, including the gables?

4. How many feet will be required to cover the whole house, including the roof, if each side of the roof is 24 ft. wide, and the waste in cutting and fitting is reckoned as equal to the space left for doors and windows?

5. What will be the cost of the boards required to cover the house at $21.50 per M (per thousand feet)?

6. If each clapboard covers a space 4 ft. long and 3 in. wide, how many clapboards will it take to cover one end of the house, including the gable?

114

Decimal Fractions

1. Multiply .5 by .3.

One tenth of one tenth is one hundredth. One tenth of five tenths is five hundredths. Three tenths of five tenths is fifteen hundredths. $.5 \times .3 = .15$. Since tenths multiplied by tenths will produce hundredths, if there are tenths in both the multiplicand and multiplier, there will be hundredths in the product.

Multiply :

2. 7 by .2   **4.** 3.5 by 7.4   **6.** 27.3 by 52.8

3. 3 by .6   **5.** 6.8 by 5.2   **7.** 46.1 by 83.9

8. 12.3 by 10.4   **10.** 38.8 by 10.8   **12.** 253.4 by 796.2

9. 20.8 by 40.1   **11.** 67.9 by 83.7   **13.** 932.8 by 600.7

14. 44.03 by 23   **16.** 37.29 by 32   **18.** 84.32 by 275

15. 59.21 by 35   **17.** 53.08 by 67   **19.** 39.41 by 846·

20. Divide 9.48 by 3.16.

$$\begin{array}{r} 3 \\ \hline 3.16)\overline{9.48} \\ 948 \\ \hline \end{array}$$

9.48 equals 948 hundredths. 3.16 equals 316 hundredths. 316 hundredths is contained in 948 hundredths 3 times.

21. Divide 25.2 by 5.04.

$$\begin{array}{r} 5 \\ \hline 5.04)\overline{25.20} \\ 2520 \\ \hline \end{array}$$

25.2 equals 252 tenths or 2520 hundredths. 504 hundredths is contained in 2520 hundredths 5 times.

Divide :

22. 8.8 by 2.2   **24.** 4.08 by 2.04   **26.** 25.35 by 5.07

23. 10.4 by 5.2   **25.** 12.18 by 4.06   **27.** 48.54 by 8.09

28. 8.20 by 2.05   **30.** 16.5 by 8.25   **32.** 32.4 by 4.05

29. 10.60 by 2.12   **31.** 24.3 by 4.05   **33.** 50.4 by 10.08

Miscellaneous Problems

1. A boy having ½ of an apple gave away ⅛ of what he had. What part of the whole apple did he give away?

2. If 3 oranges are divided among 9 persons, what part of an orange will each receive?

3. If I have 4 oranges and give ⅔ of an orange to each person, among how many persons do I divide them?

4 oranges = how many thirds?

4. How much cloth at a dollar and a half a yard can be bought for seven dollars and a half?

5. How many yards of cloth ¾ of a yard wide will line 5 yards of cloth 1½ yards wide?

The cloth is how many times as wide as the lining?

6. If cloth ¾ of a yard wide costs 75 cents a yard, what should be the price of the same quality of cloth 1¼ yards wide?

7. If cloth ⅔ of a yard wide costs 64 cents a yard, how much should the same kind of cloth cost ⅞ of a yard wide?

8. How much will 10 lbs. 4 oz. of sugar cost at 5½ cents a pound?

9. How much will 4 bu. 3 pk. of potatoes cost at ¾ of a dollar a bushel?

10. If I step two feet at each step, how many steps shall I take in going ¼ of a mile?

11. If 21 is ⅜ of some number, what is ¼ of the same number?

12. If it would take $\frac{5}{12}$ of my money to buy two neckties at 35 cents each, how much money have I?

13. After spending $\frac{7}{10}$ of my money, I had enough left to buy a bag of flour at 65 cents and a pound of butter at 28 cents. How much money had I at first?

116

Circles

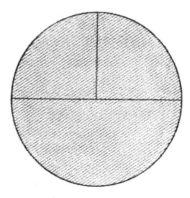

A circle is a plane figure bounded by a curved line, all points of which are at equal distances from the center within. The curved line is called the **circumference**. The distance across the circle through the center is called the **diameter**. The distance from the center to the circumference is called the **radius**. The circumference of a circle is about 3⅐, or 3.14 times the diameter.

In these problems use the number 3⅐.

1. How may the circumference of a circle be found if the diameter is given?

2. How may the circumference be found if the radius is given?

3. How may the diameter be found if the circumference is given?

4. How may the radius be found if the circumference is given?

5. If the circumference of a circle is 22 in., how long is the diameter?

6. If the circumference of a circle is 22 in., how long is the radius?

7. A wheel is 28 in. in diameter. How many feet will it move forward in turning round once on the road?

Mental Problems

1. If $\frac{1}{6}$ of a barrel of flour costs $4, what is the cost of a whole barrel?

2. If $\frac{7}{8}$ of a ton of coal costs $3.50, what will be the cost of $2\frac{1}{2}$ tons?

3. What is $\frac{1}{5}$ of 3? $\frac{1}{5}$ of 7? $\frac{1}{12}$ of 5? $\frac{7}{12}$ of 5?

One fifth of 1 is 1 fifth. One fifth of 3 is 3 fifths. One twelfth of 5 is 5 twelfths. 7 twelfths of 5 is 35 twelfths or $2\frac{11}{12}$.

4. What part of a dollar is $\frac{1}{8}$ of $5? $\frac{2}{5}$ of $2? $\frac{3}{10}$ of $3?

5. If a family will consume 2 bushels of potatoes in 3 months, what part of a bushel will they consume in 1 month?

6. If a family will consume 2 bushels of potatoes in 3 months, how many bushels will they consume in 6 months?

7. If 3 men can do a piece of work in 2 days, how long will it take 1 man to do it?

8. If 3 men can do a piece of work in 2 days, how long will it take 6 men to do it?

9. If 3 men can do a piece of work in a day, what part of it can 1 man do in a day? In half a day?

10. A merchant sold cloth for $\frac{4}{5}$ of the cost. He lost 6 cents a yard. What was the cost?

11. A merchant sold cloth for $\frac{1}{4}$ more than the cost, and gained $4\frac{1}{2}$ cents a yard. How much did he get per yard?

12. If I buy a horse for $100, keep him 6 weeks at an expense of $2.25 a week, and then sell him for $110, how much do I gain or lose?

13. A man bought a barrel of oil containing 52 gallons for $4. One fourth of the oil leaked out, and he sold the remainder at 10 cents a gallon. Did he gain or lose, and how much?

118

Original Problems

Make problems and solve them:

1. A milkman bought $12\frac{3}{4}$ gallons of milk at 16 cents a gallon.

2. A rectangular lot is 15 yards long and 40 feet wide.

3. A floor 18 ft. long and 12 ft. wide is covered with tiles.

4. A man lost $\frac{1}{4}$ of the cost by selling a horse for $60.

5. A man sold some oranges for $\frac{1}{4}$ more than the cost and received 20 cents.

6. A carpet is 18 ft. long and $13\frac{1}{2}$ ft. wide.

7. A house is built 36 ft. long, 24 ft. wide, and 18 ft. high to the eaves.

8. The roof extends 16 feet above the eaves.

9. Each side of the roof is 22 feet wide.

10. The boards cost $20 per thousand feet.

11. The clapboards cost $60 per M.

12. The diameter of a circle is 7 feet.

13. The radius of a circle is 7 feet.

14. The diameter of a wheel is 28 inches.

15. The wheel turns at the rate of 120 times a minute.

16. A man sold a horse for $80.

17. $\frac{3}{8}$ of a barrel of flour costs $4.00.

18. A family consumed 3 bushels of potatoes in 4 months.

19. 4 men can do a piece of work in 3 days.

20. $\frac{5}{8}$ of a ton of coal costs $3.00.

21. A man spent $\frac{2}{5}$ of his money.

22. A man spent $\frac{5}{8}$ of his money and had $12 left.

119

Division of Fractions

To divide fractions, when the division cannot be easily performed by inspection, first make them similar.

1. Divide $\frac{11}{12}$ by $\frac{2}{3}$.

$$\tfrac{11}{12} \div \tfrac{2}{3} = \tfrac{11}{12} \div \tfrac{8}{12} = 1\tfrac{3}{8}.$$

$\frac{8}{12}$ is contained in $\frac{11}{12}$ the same number of times that 8 is contained in 11, which is $1\frac{3}{8}$ times.

To divide mixed numbers, when the division cannot be easily performed by inspection, first change the mixed numbers to improper fractions and make the fractions similar.

2. Divide $5\frac{2}{3}$ by $2\frac{1}{5}$.

$$5\tfrac{2}{3} \div 2\tfrac{1}{5} = \tfrac{17}{3} \div \tfrac{11}{5} = \tfrac{85}{15} \div \tfrac{33}{15} = 2\tfrac{19}{33}.$$

If either the dividend or the divisor is a whole number, the whole number may be changed to the form of a fraction.

3. Divide $7\frac{3}{4}$ by 3.

$$7\tfrac{3}{4} \div 3 = \tfrac{31}{4} \div 3. \quad \tfrac{31}{4} \div \tfrac{12}{4} = 2\tfrac{7}{12}.$$

Divide :

4. $\frac{5}{8}$ by $\frac{1}{2}$.

5. $\frac{9}{10}$ by $\frac{1}{5}$.

6. $\frac{1}{2}$ by $\frac{1}{6}$.

7. $\frac{7}{12}$ by $\frac{1}{4}$.

8. $\frac{3}{4}$ by $\frac{1}{3}$.

9. $\frac{5}{7}$ by $\frac{1}{2}$.

10. $\frac{4}{5}$ by $\frac{1}{3}$.

11. $\frac{8}{9}$ by $\frac{5}{6}$.

12. $\frac{7}{8}$ by $\frac{7}{16}$.

13. $\frac{4}{5}$ by $\frac{3}{4}$.

14. $8\frac{1}{2}$ by $2\frac{1}{4}$.

15. $10\frac{5}{8}$ by $3\frac{1}{2}$.

16. $12\frac{1}{9}$ by $\frac{2}{3}$.

17. $16\frac{1}{2}$ by $1\frac{1}{7}$.

18. $18\frac{3}{10}$ by $2\frac{1}{5}$.

19. 25 by $2\frac{1}{5}$.

20. $17\frac{2}{3}$ by 5.

21. 32 by $3\frac{1}{3}$.

22. $36\frac{3}{4}$ by 4.

23. 42 by $7\frac{1}{2}$.

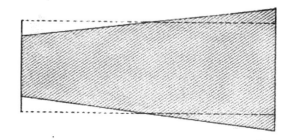

If a board is wider at one end than at the other, its area is the same as that of another board whose width is equal to the average between the two ends.

1. What is the average width of a board that is 12 inches wide at one end and 16 inches wide at the other end?

2. How many feet of lumber are there in a board that is 15 inches wide at one end, 9 inches at the other end, and 18 feet long?

3. How many feet of lumber are there in a 3-inch plank that is 2 feet wide at one end, one foot wide at the other end, and 14½ feet long?

4. If a wheel 12 feet in circumference turns round 4322 times in going a certain distance, what is the distance?

5. How many times will a wheel 13 feet in circumference turn round in going 100 miles?

6. The distance from the center of a wheel to its circumference is 21 inches. How long is the tire?

7. How many times will this wheel turn round in going 1000 miles?

8. If a circular flower bed has a diameter of 7 feet, how many plants will it take to go around the edge, if they are set 8 inches apart?

9. How many plants can be set 6 inches apart around a flower bed whose diameter is 14 feet?

121

Area of Circles

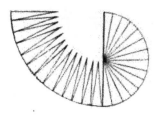

See pages 85 and 117.

A circle might be regarded as composed of a large number of small triangles, each triangle having a section of the circumference for its base, and having its vertex at the center of the circle. The altitude of the triangles is the radius of the circle.

1 What is the area of a triangle whose base is 4 in. and whose altitude is 6 in. ?

2. How might the area of two triangles be found at once, if their bases are in a continuous line and they have a common altitude ?

3. What is the combined area of three triangles, if the base of each is 3 in. long and the common altitude is 4 in. ?

4. What is the combined area of a large number of triangles, the sum of whose bases is 42 ft. and whose altitude is 10 ft. ?

5. Find the area of a circle whose circumference is 22 in. and whose radius is $3\frac{1}{2}$ in.

6. Find the circumference and the area of a circle whose diameter is 14 ft.

7. What is the diameter of a circle whose circumference is $12\frac{4}{7}$ ft. ?

8. What is the area of this circle ?

9. How many plants can be set in a circular flower bed 7 ft. in diameter, if each plant is to occupy about 2 sq. ft. ?

Triangles

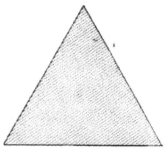
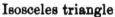
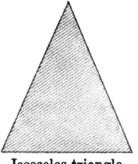

Isosceles triangle   Equilateral triangle

See pages 71 and 78.

A triangle, two of whose sides are equal to each other, is called an **isosceles triangle**. The angles which are opposite the equal sides are equal to each other.

A triangle whose three sides are all equal is called an **equilateral triangle**. The three angles are also equal.

1. How many degrees are there in the sum of all the angles of any triangle?

2. If one of the angles of a triangle is an angle of 40 degrees, how many degrees are there in the sum of the other two angles?

3. In an isosceles triangle, if the angle at the vertex is 30°, how large is one of the angles at the base?

4. In an isosceles triangle, if one of the angles at the base is 50°, how large is the other angle at the base?

5. If one of the angles at the base of an isosceles triangle is 70°, how large is the angle at the vertex?

6. How many degrees are there in the sum of all the angles of an equilateral triangle?

7. How many degrees are there in any one of the angles of an equilateral triangle?

8. If one of the angles at the base of an isosceles triangle is 60°, how large is the angle at the vertex?

123

Problems from Geography

1. If a westerly wind is blowing at the rate of 15 miles an hour, how long at that rate will it take a body of air to move from Chicago to New York, a distance of about 750 miles?

2. When a storm is moving up the coast at the rate of 40 miles an hour, how long will it take a storm which is reported to be central in Florida, to reach the New England coast, a distance of about 1200 miles?

3. How long would it take a storm center to move entirely across the country from the Pacific coast, a distance of about 2800 miles, at the rate of 25 miles an hour?

4. If a storm wind at sea were blowing at the rate of 75 miles an hour, how long would it take a body of air to move entirely across the Atlantic from Europe to the American coast, a distance of about 3000 miles?

5. The waves in the open sea sometimes move as fast as a mile a minute. If a severe storm should occur 1000 miles out at sea, how long might it be before the waves produced would reach the coast?

6. The water of the Gulf Stream moves at the rate of about two miles an hour. How long would it take for a floating piece of wood to be carried by the Gulf Stream from Florida to the coast of Europe, a distance of about 4500 miles?

7. If a ship which can usually sail at the rate of 12 miles an hour is able to get into the Gulf Stream for 5 days on its passage to Europe, how many miles of the journey will it be helped along?

8. How much will the time of the voyage be shortened by this means?

Carpeting

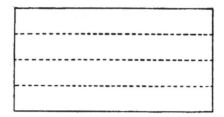

1. If a room which is 4 yards wide is to be carpeted with carpet which is 1 yard wide, and the carpet is to run lengthwise in the room, how many breadths of carpet will it take to cover the room? How many will it take to cover a room 3⅔ yards wide?

The latter room will require as many breadths as the former, the extra third yard being turned in or cut off.

2. If a room is 7 yards long and 5 yards wide, and the carpet which is 1 yard wide is to run lengthwise, how many yards of carpet will be required? How many if the carpet runs crosswise?

In the first case 5 breadths will be required, each 7 yards long; in the second case 7 breadths, each 5 yards long.

3. How many yards of carpet 1 yard wide will it take to carpet a room which is 12 feet wide and 18 feet long?

4. How many yards 1 yard wide will it take for a room 14 feet wide and 15 feet long, if the strips run lengthwise?

14 ft. ÷ 3 ft. = 4 +. 5 breadths each 15 feet long will be required.

5. What part of a yard is 27 inches?

6. How many times is ¾ contained in 4½?

7. How many breadths of carpet ¾ yd. wide will it take to extend across the width of a room 4½ yd. wide?

8. How many yards of carpet ¾ of a yard wide will be required to carpet a room 6 yards wide and 8 yards long, if the strips run lengthwise?

125

Construction

Review pages 78, 85, 92, and 123.

1. Construct a right triangle.

2. Construct a triangle with an angle of 25° and another angle of 108°.

3. Construct a parallelogram with an angle of 42°. Make a dotted line representing its altitude. Find the area.

4. Draw a horizontal line 2 inches long. Above the line, at each end, make an angle of 60°. Prolong the sides until they meet. What kind of a triangle is this? What should be the size of the third angle? Measure it. Draw a dotted line connecting this angle with the middle point of the base line. Measure the angles formed at the base by this line. Measure the altitude of the triangle. Find the area of the triangle.

5. Construct an isosceles triangle with an angle of 90° at the vertex, and with the two equal sides each 3 inches long. What should be the size of each of the other angles? Draw a dotted line for the altitude. Find the area.

6. Construct an isosceles triangle with the base line 3½ inches long.

7. Construct a triangle having a side 3 inches long and two angles of 60° each. What should be the size of the third angle? Measure the three sides. Draw a dotted line connecting one angle with the middle point of the side opposite. Measure the angles formed by this line with the side. Find the area of the triangle.

8. Construct an equilateral triangle with a side 2½ inches long.

126

Powers and Roots

The **square** of a number is obtained by multiplying two equal numbers together. The square of 3 is 3 × 3, or 9.

The **cube** of a number is obtained by multiplying three equal numbers together. The cube of 3 is 3 × 3 × 3, or 27.

The **factors** of a number are the numbers which will produce the given number when multiplied together.

1. What are two factors of 14? Two factors of 9? Three factors of 12?

The **square root** of a number is one of its two equal factors.

2. What is the square root of 4? The square root of 9? The square root of 25? The square root of 49?

The **cube root** of a number is one of the three equal factors of the number.

3. What is the cube root of 8? The cube root of 27? The cube root of 125?

4. What is the square of 6? The square root of 81?

5. What is the cube of 5? The cube root of 64?

6. What is the length of one side of a square which contains 49 square inches?

7. How long is a cube which contains 27 cubic inches?

8. The product of two equal numbers is 144. What are the numbers?

9. The product of three equal numbers is 125. What are the numbers?

10. What is the length of one side of a square field which contains 400 square rods?

11. What is the length of one edge of a cubical block which contains 216 cubic inches?

12. How many acres are there in a field which is 25 rods square?

Reduction of Fractions

A fraction is in its smallest or **lowest terms** when there is no number greater than 1 that will be contained in both the numerator and the denominator without a remainder.

To change a fraction to its lowest terms, either divide both the numerator and the denominator by the largest number that will be exactly contained in each of them, or divide both terms by any number that will be exactly contained in them, and then divide the new terms again by some other number, continuing the process until the terms can no longer be exactly divided by any number.

1. Change $\frac{16}{20}$ to its smallest terms.

$\frac{16}{20} = \frac{4}{5}$. Dividing both the numerator and the denominator by 4, we have $\frac{4}{5}$.

2. Change $\frac{36}{48}$ to its smallest terms.

$$\frac{36}{48} = \frac{3}{4}, \text{ or } \frac{36}{48} = \frac{18}{24} = \frac{9}{12} = \frac{3}{4}.$$

Dividing both terms by 12, we have $\frac{3}{4}$, or dividing by 2, 2, and 3 successively, we obtain the same result.

Change to their lowest terms:

3. $\frac{8}{12}$.	13. $\frac{32}{48}$.	23. $\frac{60}{72}$.	33. $\frac{26}{34}$.
4. $\frac{16}{24}$.	14. $\frac{27}{36}$.	24. $\frac{42}{76}$.	34. $\frac{75}{85}$.
5. $\frac{20}{36}$.	15. $\frac{48}{64}$.	25. $\frac{35}{55}$.	35. $\frac{16}{80}$.
6. $\frac{30}{50}$.	16. $\frac{24}{40}$.	26. $\frac{60}{75}$.	36. $\frac{66}{88}$.
7. $\frac{48}{56}$.	17. $\frac{35}{63}$.	27. $\frac{54}{81}$.	37. $\frac{32}{100}$.
8. $\frac{18}{54}$.	18. $\frac{36}{72}$.	28. $\frac{68}{84}$.	38. $\frac{36}{96}$.
9. $\frac{20}{45}$.	19. $\frac{45}{60}$.	29. $\frac{55}{66}$.	39. $\frac{55}{80}$.
10. $\frac{25}{75}$.	20. $\frac{32}{64}$.	30. $\frac{44}{76}$.	40. $\frac{80}{95}$.
11. $\frac{35}{80}$.	21. $\frac{40}{75}$.	31. $\frac{56}{84}$.	41. $\frac{75}{100}$.
12. $\frac{30}{48}$.	22. $\frac{27}{72}$.	32. $\frac{45}{90}$.	42. $\frac{50}{155}$.

English Money

Twenty **shillings** of English money make a **pound.**
Twelve **pence** make a **shilling.**

1. How many shillings are there in $3\frac{1}{4}$ pounds?
2. How many pence make a pound?
3. How many shillings in £7 4s.?
4. How many pence in 4s. 7d.?
5. How many pence in £2 10s. 6d.?
6. Add £3 5s. 2d.; £5 7s. 4d.; and £9 8s. 6d.
7. From £10 5s. 11d. take £4 7s. 5d.
8. Multiply £3 9s. 3d. by 5.
9. Divide £8 6s. 3d. by 7.

10. If a book costs 10s. 6d., how many books of the same kind can be bought for £2 12s. 6d.?

11. If a book costs 8s. 4d., how much will 20 books of the same kind cost?

12. A pound of English money is equivalent to about $4.86 of United States money. A shilling of English money is equivalent to about 25 cents of United States money. A penny of English money is equivalent to about 2 cents of United States money. What would be the equivalent in United States money of £4 3s. 3d.?

13. About how much United States currency would be equivalent to £7 8s. 4d.?

14. How much English money is about equal to $10.75?

15. About how much English money could be obtained in exchange for $12.33?

16. If cloth in London costs 7s. per yard, what will be the whole cost when imported, if the freight amounts to 5 cents per yard and the import duty is 50 cents per yard?

Review Problems

1. A house is 70 ft. long, 36 ft. wide, and 17 ft. high to the eaves. The roof extends 16 ft. above the line of the eaves. How many square feet are there in the two gables?

2. How much will it cost to paint the entire surface of the walls of the house, including the gables, at 12½ cents a square yard?

3. How much lumber is there in a plank 12 ft. long, 8¼ in. wide at one end and 11¾ in. wide at the other end, and 2 in. thick?

4. It is exactly a mile across the center of a circular pond. What is the distance around it?

5. How many rods is it around a circular race course, if the distance from the fence to the center is 56 rods?

6. One of the acute angles of a right triangle is 35°. What is the size of the other acute angle?

7. One of the equal angles of an isosceles triangle is 40°. How large is the angle at the vertex?

8. How many yards of carpet will it take to carpet a room which is 21 ft. long and 15 ft. wide, if the carpet is 1 yd. wide and the breadths run lengthwise?

9. Find the square of 18; 250; 1000.

10. Find the cube of 25; 50; 100.

11. What is the length of a square which contains 100 sq. in.?

12. What is the length of a cube which contains 216 cubic in.?

13. To about how much United States money is £12 7 *s*. 11 *d*. equivalent?

Drill Work

Add the following and prove the work:

1.	2.	3.	4.
$197.24	$327.81	$356.75	$869.85
46.82	117.94	61.37	18.21
87.36	64.67	569.50	75.36
497.22	42.80	81.86	847.60
863.14	383.62	241.11	627.71
304.41	738.76	935.60	4.62
7.97	9.19	76.13	491.40
76.90	13.33	439.07	75.35
684.18	497.66	9.65	46.82
220.36	286.10	276.97	986.25
79.65	12.64	65.52	761.97
61.51	856.05	344.04	83.47
869.85	48.17	7.87	58.49
726.29	939.93	753.50	532.10

5.	6.	7.	8.
$641.52	$698.93	$197.19	$684.60
92.14	784.74	263.04	45.28
576.32	99.48	74.97	52.43
96.77	6.76	9.72	767.34
41.61	565.49	352.25	485.97
8.20	647.50	968.08	38.50
669.89	67.92	50.33	4.86
714.21	983.29	534.36	79.83
84.69	474.77	440.50	196.02
732.65	9.56	7.88	643.41
529.50	44.40	824.90	259.93
64.78	836.59	675.04	93.07
123.25	58.24	384.50	627.35
6.89	597.95	76.97	874.24

Falling Bodies

When a body falls toward the earth it continually increases in speed while falling. A heavy body will fall faster than a light body because the speed of the light body is checked more by the air in falling through it. A heavy object like a stone will fall about 16 feet in one second, but in the following second it will fall more than 16 feet because of the constant increase in speed.

To find how far a heavy object will fall in any number of seconds square the number indicating the number of seconds and multiply this square by 16. In 2 seconds an object will fall about 2 × 2 × 16 feet; in 3 seconds 3 × 3 × 16 feet, and so on.

1. If a stone should be dropped into a well, and should reach the bottom in exactly one second, how deep must the well be?

2. Find how deep the well would be if the stone were 2 seconds in falling.

3. A stone dropped from a cliff reaches the base in 4 seconds. What is the height of the cliff?

The speed of a stone which is thrown directly upward becomes gradually less until it stops. The stone is precisely as long in going up as in coming down.

4. If a stone thrown directly upward reaches the ground again in exactly 2 seconds, how high did it rise?

5. If a stone reaches the ground in 6 seconds after having been thrown upward, how many feet did it rise?

6. If a stone is dropped from a high cliff into the water and the sound of the splash is heard upon the cliff in 5½ seconds, how high is the cliff, if ½ a second is allowed for the sound of the splash to ascend from the water?

Original Problems

Make problems and solve them:

1. A board is 15 inches wide at one end and 22 inches wide at the other.

2. A board is 10 inches wide at one end, 16 inches wide at the other, and 8 feet long.

3. The distance from the center of a wheel to its circumference is 28 inches.

4. The diameter of a flower bed is 14 feet.

5. One of the equal angles of an isosceles triangle is 40°.

6. The angle at the vertex of an isosceles triangle is 24°.

7. The Gulf Stream moves at the rate of about 2 miles an hour.

8. A room 18 ft. square is to be carpeted.

9. A room 21 ft. long and 18 ft. wide is to be carpeted.

10. A room is to be carpeted with carpet $\frac{3}{4}$ of a yard wide.

11. The floor of a certain square room contains 64 square yards.

12. A certain square field contains 900 square rods.

13. A cubical block contains 125 cubic inches.

14. An American traveler bought 4 books in England for 7*s.* 6*d.* each.

15. A traveler exchanged $40 for English money.

16. A certain square field is 100 rods square.

17. A house is 24 ft. wide, and the roof extends 18 ft. above the eaves.

18. A stone dropped from a cliff over the water reaches the water in 3 seconds.

Decimal Fractions

Millions	Hundred thousands	Ten thousands	Thousands	Hundreds	Tens	Units	Tenths	Hundredths	Thousandths	Ten-thousandths	Hundred-thousandths	Millionths
2	3	0	7	5	6	4.	5	1	0	7	3	5

In a fraction written in the decimal form the first figure to the right of the decimal point indicates the number of *tenths*, the second figure *hundredths*, the third *thousandths*, and so on. The last figure gives the name to the number. .006 is six *thousandths*. .0132 is one hundred thirty-two *ten-thousandths*.

Read the following numbers:

1.	2.	3.	4.
.8	.0004	.2533	45.023
.05	.0015	.253	156.17
.006	.0126	.25	5.1356

5. Write: twenty-five and six hundredths; eight and three ten-thousandths; one hundred and three thousandths; one hundred ten and two hundredths.

6. Find the sum of five and twenty-four hundredths, seven and eight tenths, five and sixty-four hundredths, twenty-two and five hundredths.

7. Find the sum of eighteen and seven thousandths, twenty-five and twenty-five thousandths, three hundred seventy-five thousandths, nine and twenty-six hundredths.

8. From three and eight hundred fifty-six thousandths take sixty-seven hundredths.

9. From six thousandths take eight ten-thousandths.

134

Review of Measures

1. If 56 tons of hay cost $1344, what is the price per ton?

2. At the rate of $1.15 per hundred pounds, how much will 23¼ tons of hay cost?

3. What part of 25 is 4? Of 75 is 45?

4. What part of a cubic yard is 3 cubic feet?

5. What part of a square foot is an oblong 6 in. long and 2 in. wide?

6. What is the cost of 14 qt. of apples at 28 cents a pk.?

7. Mt. Shasta is 14,440 ft. high. How many miles high is it?

8. How many square feet of boards will be needed to make a box 8 ft. long, 6 ft. 6 in. wide, and 4 ft. deep, if the waste equals the difference due to overlapping at the edges?

9. How many 3-inch cubes are there in a block 9 in. long, 6 in. wide, and 3 in. thick?

10. How many yards of carpet 1 yard wide will it take to cover a floor which is 27 ft. long and 5 yd. wide, if no allowance is made for waste?

11. How many rods of fence will be required to build a fence on both sides of a railroad 15 miles long?

12. How many square yards are there in the four walls of a room 15 ft. long, 12 ft. wide, and 9 ft. high?

13. A rectangle 4 yd. long contains 60 sq. ft. How wide is it?

14. A rectangular solid 7 ft. wide and 3 ft. thick contains 210 cubic feet. How long is it?

15. How many 3-in. cubes will it take to make a 9-in. cube?

135

Fractions — Drill Work

Change to whole numbers or mixed numbers:

1. $\frac{65}{6}$.
2. $\frac{84}{9}$.
3. $\frac{78}{8}$.
4. $\frac{69}{10}$.
5. $\frac{84}{12}$.
6. $\frac{95}{9}$.
7. $\frac{98}{11}$.

8. $\frac{100}{12}$.
9. $\frac{120}{9}$.
10. $\frac{145}{8}$.
11. $\frac{157}{11}$.
12. $\frac{161}{7}$.
13. $\frac{184}{9}$.
14. $\frac{197}{12}$.

15. $\frac{215}{18}$.
16. $\frac{228}{15}$.
17. $\frac{248}{17}$.
18. $\frac{200}{16}$.
19. $\frac{285}{20}$.
20. $\frac{324}{18}$.
21. $\frac{395}{25}$.

22. $\frac{400}{23}$.
23. $\frac{464}{27}$.
24. $\frac{510}{30}$.
25. $\frac{632}{35}$.
26. $\frac{850}{40}$.
27. $\frac{975}{25}$.
28. $\frac{1000}{45}$.

Change to improper fractions:

29. $5\frac{7}{8}$.
30. $9\frac{9}{10}$.
31. $7\frac{7}{16}$.
32. $8\frac{4}{11}$.
33. $4\frac{15}{16}$.
34. $6\frac{18}{19}$.
35. $9\frac{17}{20}$.

36. $12\frac{4}{9}$.
37. $15\frac{9}{10}$.
38. $18\frac{5}{18}$.
39. $20\frac{13}{17}$.
40. $25\frac{9}{20}$.
41. $28\frac{7}{25}$.
42. $30\frac{11}{18}$.

43. $22\frac{11}{12}$.
44. $29\frac{7}{16}$.
45. $31\frac{3}{10}$.
46. $36\frac{7}{20}$.
47. $37\frac{3}{17}$.
48. $40\frac{4}{23}$.
49. $45\frac{5}{27}$.

50. $34\frac{7}{8}$.
51. $42\frac{3}{11}$.
52. $48\frac{5}{22}$.
53. $54\frac{29}{30}$.
54. $87\frac{13}{36}$.
55. $120\frac{7}{18}$.
56. $245\frac{39}{40}$.

Change to improper fractions and divide:

57. $7\frac{2}{3} \div 2\frac{1}{2}$.
58. $9\frac{3}{5} \div 1\frac{7}{10}$.
59. $12\frac{7}{8} \div \frac{3}{4}$.
60. $10\frac{1}{3} \div 3\frac{2}{5}$.
61. $15\frac{6}{7} \div 4\frac{1}{2}$.
62. $18\frac{2}{9} \div \frac{5}{6}$.
63. $21\frac{3}{7} \div 4$.

64. $25 \div \frac{2}{3}$.
65. $28 \div 2\frac{3}{5}$.
66. $36\frac{1}{9} \div 5$.
67. $42\frac{3}{4} \div 12$.
68. $32\frac{5}{9} \div 1\frac{1}{6}$.
69. $47\frac{11}{12} \div 8\frac{3}{4}$.
70. $40\frac{9}{10} \div \frac{9}{10}$.

71. $53\frac{1}{2} \div 17\frac{1}{3}$.
72. $56 \div \frac{12}{13}$.
73. $\frac{25}{28} \div 4$.
74. $10 \div \frac{7}{32}$.
75. $125\frac{1}{2} \div 12\frac{1}{3}$.
76. $156\frac{2}{3} \div 14\frac{5}{6}$.
77. $225\frac{7}{12} \div 114\frac{3}{4}$.

136

Mentel Problems

Illustrate by diagrams. See pages 61 and 93.

1. Divide the number 8 into 2 parts, one of which shall be 3 times as large as the other.

Find $\frac{1}{4}$ of 8 and $\frac{3}{4}$ of 8. See page 93.

2. Divide the number 15 into 2 parts, one of which shall be twice as large as the other.

3. Divide the number 12 into 2 such parts that one shall be five times the other.

4. Divide the number 18 into 2 such parts that one shall be 8 times the other.

5. How may 12 cents be divided between 2 boys so that one may have 3 times as many as the other?

6. Charles and William together have 14 cents. If Charles has 6 times as many as William, how many cents has each?

7. 3 children together have 21 cents. If the second has twice as many as the first and the third twice as many as the second, how much money has each?

Once the first child's money plus twice his money plus 4 times his money or 7 times his money equals 21 cents.

8. John, Mary, and Anna together have 50 cents. John has 5 times as many as Mary, and Anna has 4 times as many as Mary. How many has each?

9. I have twice as many dimes as 5-cent pieces and three times as many cents as 5-cent pieces. I have 20 dimes. How many coins have I in all?

10. A farmer has $\frac{1}{2}$ as many turkeys as ducks and 4 times as many hens as turkeys. He has 40 hens. How many has he in all?

137

1. How many sheets of paper are there in a quire?

2. How many quires in a ream?

3. How many single sheets are there in a ream of paper?

4. How many sheets are there in 3 reams and 6 quires?

5. How many sheets are there in $4\frac{1}{2}$ reams?

6. I bought 5 reams of paper at $1.25 a ream and sold it at 10 cents a quire. How much did I gain?

7. If I should buy a ream of paper for $2.50, and sell it for a cent a sheet, how much should I make?

8. If I should buy $\frac{1}{2}$ a ream of paper for 75 cents, and sell it for 10 cents a quire, how much should I make?

9. If I should buy 2 reams of paper at the rate of 8 cents a quire, and should sell it at the rate of 3 sheets for 2 cents, what would my profit be?

10. If envelopes are put up 25 in a package, and I reckon 2 sheets of paper for each envelope, how many packages of envelopes should I buy with 3 reams of paper?

11. If 500 envelopes cost 80 cents, how much is that per package of 25?

12. If I should buy a ream of paper for $1.25, and 20 packages of envelopes at 4 cents a package, and should sell the paper for a cent a sheet, and the envelopes for a cent each, how much profit should I make?

13. Find the profit upon 2 reams of paper bought at $1.15 per ream, if one half of it is sold at 8 cents a quire and the remainder at a cent a sheet.

138

The Thermometer

In a common thermometer the mercury in the tube expands and rises when the temperature becomes warmer, and contracts and falls when the temperature becomes colder. The tube is marked with spaces called **degrees**. In the Fahrenheit thermometer, which is most commonly used, the point at which water freezes is marked 32°, and the point at which water boils is marked 212°.

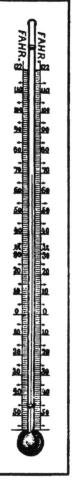

1. On the Fahrenheit thermometer how many degrees are there from the freezing point to the boiling point?

2. How many degrees are there from the freezing point to 30° below zero?

3. If at 2 o'clock P.M. the mercury is at $92\frac{1}{2}$°, and is steadily falling at the rate of $6\frac{1}{2}$° an hour, at what time will it reach 60°?

4. If water at the freezing point is heated over a fire, and its temperature is rising at the rate of 9° a minute, how long will it be before it will boil?

5. From 3 o'clock P.M. to 8.30 P.M. the mercury fell 22°. At what average rate per hour did the mercury fall?

6. The thermometer at 8 o'clock in the evening on 7 successive days indicated as follows: 62°, $53\frac{1}{2}$°, $56\frac{1}{2}$°, 64°, $61\frac{1}{2}$°, $52\frac{1}{2}$°, 55°. What was the average temperature for the week?

7. For 5 mornings in succession the mercury stood as follows: −10°, −7°, −3°, −4°, −5°. What was the average temperature for the 5 mornings?

Review Problems

See pages 70 and 84.

1. Find the exact number of days from Jan. 1st to Feb. 7th.

2. How many days are there from Washington's Birthday to Independence Day?

3. If the first day of June is Saturday, what day of the week will the first day of July be?

4. If the tenth of August is Tuesday, what day of the week will the fifth of September be?

5. How many even months and how many days over are there from May 30th to Dec. 15th?

6. How many even months and how many days over from March 12th to Nov. 18th?

7. Find the time from June 12, 1843, to Sept. 25, 1859.

8. Find the time from Oct. 10, 1887, to March 16, 1895.

9. Find the time from Dec. 30, 1873, to Jan. 1, 1885.

10. How far will a heavy body fall in 3 seconds?

See page 132.

11. How long will it take a heavy body to fall 64 feet?

12. If a stone thrown directly upward reaches the ground again in 6 seconds, how high did it ascend?

13. In a clear, mild day the report of a gun is heard $2\frac{1}{2}$ seconds after the flash is seen. What is the distance of the gun?

See page 86.

14. In a cold day the whistle of a locomotive is heard 4 seconds after the steam is seen. How far distant is the locomotive?

15. How far will sound travel in water in half a second?

140

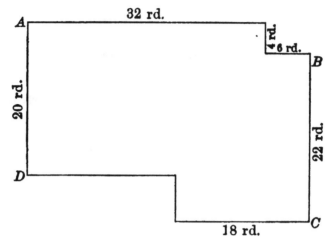

The diagram represents a lot of land.

1. Find the entire distance around it in yards.

2. Find the area of the lot in square rods.

3. How many acres does it contain?

4. If I should have one acre of this land occupied with corn, 1¼ acres with potatoes, 1½ acres with oats, and the remainder with grass, how many square rods of grass would there be?

5. How much would it cost to build a stone wall 3 ft. wide and 4½ ft. high across the west end of the lot, at 60 cents a cubic yard?

6. Which would cost the less, and how much less; to build such a wall at 60 cents a cubic yard, or at $5 for a rod in length?

7. How many rods is it from the northwest corner to the southeast corner of the lot, measuring around the line by the north side?

8. Which is the shorter distance from *A* to *C*, and how much, measuring the line by way of *B* or by way of *D*?

9. How much difference is there in the two distances from *D* to *B*, measuring by *A* and by *C*?

141

Drill Work

Add:

1.	**2.**	**3.**	**4.**
$5\frac{3}{8}$	$9\frac{1}{5}$	$10\frac{3}{8}$	$13\frac{2}{5}$
$7\frac{1}{4}$	$7\frac{3}{4}$	$12\frac{1}{6}$	$18\frac{7}{15}$
$6\frac{1}{2}$	$5\frac{7}{10}$	$15\frac{7}{12}$	$15\frac{5}{6}$

Subtract:

5.	**6.**	**7.**	**8.**
$21\frac{3}{8}$	$32\frac{1}{6}$	$24\frac{5}{8}$	$36\frac{1}{7}$
$12\frac{3}{4}$	$18\frac{7}{12}$	$10\frac{1}{3}$	$18\frac{1}{4}$

9.	**10.**	**11.**	**12.**
$56\frac{7}{15}$	$48\frac{3}{11}$	$87\frac{1}{2}$	$95\frac{1}{2}$
$42\frac{1}{6}$	$24\frac{1}{2}$	$61\frac{5}{9}$	$80\frac{3}{7}$

Multiply:

13.	**14.**	**15.**	**16.**
$9\frac{1}{2}$	$10\frac{2}{3}$	$13\frac{5}{8}$	$14\frac{8}{9}$
$8\frac{1}{4}$	$6\frac{1}{2}$	$8\frac{1}{5}$	$9\frac{1}{4}$

17.	**18.**	**19.**	**20.**
$16\frac{4}{5}$	$21\frac{9}{10}$	$42\frac{7}{12}$	$55\frac{5}{6}$
$15\frac{3}{4}$	$20\frac{2}{3}$	$18\frac{8}{7}$	$36\frac{4}{5}$

21. Divide $21\frac{1}{4}$ by 5.

In dividing, if there is a whole number and a fraction remaining, change this mixed number to the form of a fraction, and then divide.

$5\overline{)21\frac{1}{4}}$ 5 is contained in $21\frac{1}{4}$ 4 times, with $1\frac{1}{4}$ remaining. $1\frac{1}{4} \div 5$
$\quad 4\frac{1}{4}$ $= \frac{5}{4} \div 5 = \frac{1}{4}$. The answer is $4\frac{1}{4}$.

Divide:

22.	**23.**	**24.**	**25.**
$6\overline{)444\frac{6}{7}}$	$4\overline{)845\frac{3}{5}}$	$5\overline{)766\frac{3}{7}}$	$3\overline{)866\frac{1}{10}}$

26.	**27.**	**28.**	**29.**
$7\overline{)925\frac{3}{11}}$	$9\overline{)658\frac{4}{5}}$	$8\overline{)282\frac{2}{3}}$	$9\overline{)893\frac{7}{10}}$

142

Original Problems

Make problems and solve them:

1. Hay costs $1.10 per 100 pounds.

2. Apples cost 6 cents a quart.

3. A fence is to be built on both sides of a railroad 10 miles long.

4. A room is 15 ft. long, 12 ft. wide, and 9 ft. high.

5. Lead weighs 11.35 times as much as water.

6. 10 cubic feet of cork weigh 2400 ounces.

7. $\frac{1}{2}$ of a cubic foot of iron weighs 3600 ounces.

8. A box contains 2 gross of pencils.

9. A dealer bought pens for 60 cents a gross.

10. 24 cents are divided between two boys.

11. A dealer bought paper for 10 cents a quire and sold it for a cent a sheet.

12. I wrote 2 letters each day during the month of June, and used 2 sheets of paper for each letter.

13. The Declaration of Independence was adopted July 4, 1776.

14. At 3 o'clock the mercury in a Fahrenheit thermometer stood at 42°. After that it fell at the rate of a degree in 12 minutes.

15. For 4 successive evenings at 8 o'clock the temperature was $-12°$, $-4°$, $-2°$, and $-6°$.

16. The first day of April 1901 was Monday.

17. A stone fell 64 feet.

18. The whistle of a steamboat was heard 4 seconds after the steam was seen.

19. A fence is to be built around a lot of land 22 rd. long and 16 rd. wide.

Decimal Fractions

$$5\tfrac{3}{10} = 5.3 = \tfrac{53}{10}. \qquad 17\tfrac{205}{1000} = 17.205 = \tfrac{17205}{1000}.$$

$$\tfrac{7}{10} \times \tfrac{481}{100} = \tfrac{3367}{1000}. \qquad .7 \times 4.81 = 3.367.$$

Notice that the number of decimal figures in any decimal fraction is equal to the number of ciphers which are annexed to the figure 1 in the denominator of the fraction, and that *the number of decimal figures in the product is equal to the sum of the number of decimal figures in the multiplicand and the multiplier.*

1. Multiply 24.207 by .83.

2. Multiply 10.0518 by 3.181.

3. Multiply .075 by .0013.

4. Multiply 250 by .0025.

5. Multiply 5000 by .0005.

$$\tfrac{21}{100} \div \tfrac{3}{10} = \tfrac{7}{10}. \qquad .21 \div .3 = .7.$$

As division is the reverse of multiplication, the number of decimal figures in the quotient is equal to *the difference between the number in the dividend and the number in the divisor.*

6. Divide 43.50 by 4.35.

7. Divide 18.72 by 31.2.

8. Divide 8.075 by 3.23.

9. Divide 131.25315 by 1.05.

10. Divide 288.402264 by 15.72.

11. .0536 × 7.4 = ?

12. 367 × .0004 = ?

13. 46.240 ÷ 20 = ?

14. .00129 ÷ 4.3 = ?

15. 3.72812 ÷ 8.14 = ?

16. .00255 ÷ 102 = ?

17. 70.3 × .045 = ?

18. .6723 ÷ 24.9 = ?

19. 95.04 × 3.513 = ?

20. .0021318 ÷ 19 = ?

144

Review Problems

1. If one of the acute angles of a right triangle is 25°, how large must the other acute angle be ?

2. If the sum of 2 of the angles of a triangle is 99°, how large must the remaining angle be ?

3. If the three angles of a triangle are all equal, how large is each angle ?

4. If the diameter of a wheel is 42 inches, what is its circumference ?

5. If the full moon appears to the eye to be about 28 inches in diameter, how many square feet does it appear to have ?

6. How far will a stone fall in 2 seconds ?

7. If a stone is thrown to the height of 64 feet, how long is it in both ascending and descending ?

8. How many cubic feet in a block of marble 4 ft. wide, 1 ft. 3 in. thick, and 6 ft. long ?

9. If the marble weighs 2.83 times as much as water, how many ounces would this block weigh ?

10. How much more than 2 tons would the block weigh?

11. If a teacher has a school of 420 pupils, and has 6 gross of pencils, how many pencils will he have left after giving each pupil 2 pencils?

12. Find the time from Aug. 13, 1846, to June 18, 1873.

13. Find the exact number of days from Nov. 12, 1900, to May 1, 1901.

14. At what age did a person die who was born Mar. 5, 1828, and died Dec. 2, 1896 ?

15. Queen Victoria was born May 24, 1819, and died Jan. 22, 1901. What was her exact age at death ?

Integral Numbers

1. Write in figures: twenty-five thousand, four hundred fifty-two; one hundred seventy thousand, five hundred; one million, forty-two thousand, three hundred fifty; three thousand, fifty; one million, one hundred thousand, twenty-five.

2. Write in figures: five thousand, twenty-five; fifty thousand, fifty; two hundred thousand, ten; one million, one thousand, one hundred; one hundred million, one thousand, ten.

3. Find the sum of one thousand, two hundred fifty; ten thousand, one hundred, nine; five hundred thousand; sixty thousand, five hundred, ten; two million, four hundred thousand, five hundred.

4. Find the sum of one thousand, nine hundred ninety-nine; sixty thousand, five hundred; two hundred two thousand, twenty; fifteen million, nine thousand, nine; two hundred thousand, ten; forty million, twenty-five thousand, nine hundred.

5. From 5850 take 648.

6. From 12,000 take 2101.

7. From 93,000,000 take 240,000.

8. Multiply 8468 by 296.

9. Multiply 953 by 1000.

10. Multiply 62,540 by 5000.

11. Multiply 8937 by 3400.

12. Divide 186,000 by 93.

13. Divide 50,000,000 by 5000.

14. Divide 8650 by 100.

15. Divide 24,500 by 1000.

16. Divide 372,500 by 10,000.

146

Review Problems

1. The circumference of a circle is 3.1416 times its diameter. Find the circumference of a circle whose diameter is 35 inches.

2. Find to the nearest hundredth the circumference of a circle whose diameter is 27.8 inches.

3. How far is it around a circular pond whose diameter is 2.5 miles?

4. How many plants can be set not less than one foot apart around the border of a circular bed whose diameter is 7.25 feet?

5. A bicycle wheel is 26 inches in diameter. Find how many times the wheel turns in going a mile.

6. A cubic foot of water weighs 62.5 pounds. Mercury weighs 13.59 times as much as water. Find the weight of a cubic foot of mercury.

7. Find to the nearest hundredth the weight of a cubic inch of mercury.

8. A cubic foot of granite weighs 170 pounds. How many times as heavy as water is granite?

9. Find to the nearest thousandth the weight of a cubic inch of granite.

10. How many times as heavy as granite is mercury?

11. Lead is 11.35 times as heavy as water, and wrought iron is 7.77 times as heavy as water. How many times as heavy as lead is wrought iron?

12. A pound of pure gold is worth $248.04. There are 12 ounces of gold in a pound. How much is an ounce of gold worth?

13. Find how many pounds of gold it would take to be worth $1,000,000.

Problems from Geography

1. The diameter of the earth at the equator is 7926 miles. About how far is it around the earth?

2. If the circumference of the earth is regarded as divided into 360 degrees, about how many miles are there in one degree?

3. If two places near the equator are 5 degrees of longitude apart, about how many miles apart are they?

Since a circle forming a parallel of latitude north or south of the equator is not so large as the circle around the earth at the equator, one degree on such a circle is not so long as one degree on the circle at the equator.

4. At a certain latitude where the degrees of longitude are 42 miles long, how far apart are two places whose difference in longitude is $12\frac{3}{4}$ degrees?

5. Find how many miles an hour a person at the equator is carried around by the earth in its daily revolution.

6. How many times faster is a person at the equator carried around by the earth than one is carried by an express train which goes at the rate of 40 miles an hour?

7. How many days would it take to go around the earth by train and steamboat, traveling day and night, at an average rate of 20 miles an hour?

8. From the earth to the moon it is about 240,000 miles. If it were possible to go to the moon in a flying machine at the rate of 30 miles an hour, how long would it take?

9. The moon revolves around the earth in about 27 days. Its path is not quite a circle, but if we regard it as a circle, about how far does the moon travel in going around?

10. How many miles a day does the moon travel?

Miscellaneous Drill Work

1. $23.57 + 8.405 + 150.9 + .8309 = ?$
2. $.893 + 39.5 + 283.95 + 16.0416 = ?$
3. $.058 + 360 + 29.318 + 3.9 = ?$
4. $37.005 + 3.714 + 200 + .002 = ?$
5. $156.25 + 78.04 + .963 - 83.09 = ?$
6. $3.5203 + 95.4 - .1306 - 27.5 = ?$
7. $1000 - 236 - 150.8 - .89513 = ?$
8. $2500 + 250.25 + .0125 - 918.0005 = ?$

Multiply :

9. $.25 \times 16$
10. $.075 \times 42$
11. $.93 \times .27$
12. 5.83×3.9
13. 56.85×5.8
14. $.85 \times 32.7$
15. 8.62×9.43
16. $.576 \times .241$

17. $245 \times .023$
18. $836 \times .005$
19. $927 \times .001$
20. $3.064 \times .08$
21. $.007 \times .009$
22. $350 \times .035$
23. $2000 \times .002$
24. $175 \times .0008$

Divide :

25. $76.92 \div 8$
26. $145.44 \div 12$
27. $126.25 \div 25$
28. $36.55 \div 4.3$
29. $68.37 \div 5.3$
30. $2.744 \div 5.6$
31. $5.893 \div .83$
32. $437.75 \div 2.5$

33. $299.544 \div 42$
34. $3.3626 \div 7.31$
35. $6.7252 \div 46$
36. $3.036 \div .012$
37. $12.144 \div 253$
38. $.1452 \div .605$
39. $.4356 \div .24$
40. $.031812 \div .132$

Review of Measures

1. How many quarter pound packages can be put up from 15 lb. 12 oz. ?

2. How many half pint bottles can be filled from 3 gal. 2 qt. of cream ?

3. What would be the cost of 6 lb. 8 oz. of butter at 22½ cents a pound ?

4. What will be the cost of 3 pk. 2 qt. of nuts at 40 cents a peck ?

5. Find the area of a rectangle 7 yd. 2 ft. long and 9 ft. wide.

6. What is the area of a rectangle 3 rd. 12½ ft. long and 18½ ft. wide ?

7. A field containing 120 sq. rd. is 24 rd. long. How wide is it ?

8. What part of a square yard is 3 sq. ft. ?

9. What part of an acre is a piece of land 10 rd. long and 6 rd. wide ?

10. What are the contents in cubic feet of a box 6 ft. 3 in. long, 4 ft. wide, and 3½ ft. deep ?

11. How many cubic feet of earth will have to be removed to dig a cellar 45 ft. long, 32 ft. wide, and 9 ft. deep ?

12. How many square feet of galvanized iron will it take to line the sides and bottom of a box 12⅓ ft. long, 3¼ ft. wide, and 8 ft. deep ?

13. How many feet of lumber are there in a plank 24 ft. long, 9 in. wide, and 2 in. thick ?

14. How many feet of boards will it take to build a fence 10 rd. long, 4 boards high, if the boards are 6 in. wide ?

150

Circles within Squares

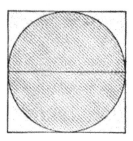

A circle is **inscribed** in a square when the center of the circle is at the center of the square and the circumference of the circle just touches the four sides of the square.

1. How long is the diameter of a circle which is inscribed in a square 28 in. long?

2. How much farther is it around the square than around the circle?

3. What is the area of the circle?

4. Find the area of the space outside the circle and within the square.

5. What fractional part of the surface of the square is the surface of the circle?

6. What fractional part of the surface of the square is left after the circle is taken from it?

7. What is the diameter of the largest circle which can be inscribed in a square containing 144 sq. in.?

What is the width of the square?

8. A circle whose radius is $6\frac{1}{2}$ in. is inscribed in a square. What is the area of the square?

9. The area of a square is 196 sq. in. Find the area of a circle inscribed in this square.

10. The circumference of a circle which is inscribed in a square is 22 ft. Find the length of the perimeter of the square.

The perimeter of a figure is its outer boundary.

Miscellaneous Drill Work

How many yards of carpet one yard wide will it take to exactly cover —

1. A floor 8 yards long and 5 yards wide?
2. A floor 10½ yards long and 6½ yards wide?
3. A floor 32 feet long and 15 feet wide?
4. A floor 13½ feet long and 4 yards wide?
5. A floor 12⅓ feet long and 27 feet wide?

How many yards of carpet ¾ of a yard wide will it take to exactly cover —

6. A floor 5 yards long and 3 yards wide?
7. A floor 7 yards long and 4½ yards wide?
8. A floor 24 feet long and 18 feet wide?
9. A floor 6 yards long and 15 feet wide?
10. A floor 8 yards long and 3¾ yards wide?

Find the measurements of a circle:

11. The circumference, if the diameter is 15 inches.
12. The circumference, if the diameter is 28.5 inches.
13. The diameter, if the circumference is 42 inches.
14. The diameter, if the circumference is 74.56 inches.
15. The circumference, if the diameter is 5½ feet.
16. The diameter, if the circumference is 17½ feet.
17. The circumference, if the radius is 8 inches.
18. The area, if the radius is 8 inches.
19. The area, if the radius is 20 inches.
20. The diameter, if the circumference is 25 feet.
21. The area, if the circumference is 25 feet.
22. The area, if the diameter is 50 feet.

152

Original Problems

Make problems and solve them :

1. The area of a right triangle is 48 square feet.

2. The altitude of a parallelogram is 15 inches. Its base is 28 inches.

3. One of the acute angles of a right triangle is 28°.

4. One angle of a triangle is 98° and another angle is 35°.

5. A stone thrown directly upward returns to the ground in 4 seconds.

6. A teacher has 5 gross of pencils.

7. 13 cubic feet of air weigh about a pound.

8. The polar diameter of the earth is 7900 miles.

9. At a certain latitude the degrees of longitude are $44\frac{1}{2}$ miles long.

10. A tourist traveled around the earth at an average rate of 12 miles an hour.

11. The distance from the earth to the moon is about 240,000 miles.

12. A box is 3 ft. 8 in. long, 2 ft. 6 in. wide, and 2 ft. deep.

13. 45 is $\frac{5}{8}$ of a certain number.

14. $1\frac{3}{4}$ acres of land cost $70.

15. A circle is inscribed in a square which is 14 inches square.

16. The circumference of a circle which is inscribed in a square is 44 inches.

17. A floor is 18 feet long and 13 feet wide.

18. A floor is to be covered with a carpet 1 yard wide.

153

1. What is $\frac{2}{5}$ of 35?

2. What is $\frac{7}{8}$ of 56?

3. 18 is $\frac{2}{5}$ of what number?

4. 42 is $\frac{7}{12}$ of what number?

5. What number is $\frac{1}{3}$ more than 9?

   $\frac{1}{3}$ more than 9 = $\frac{4}{3}$ of 9.

6. What number is $\frac{1}{6}$ less than 24?

   $\frac{1}{6}$ less than 24 = $\frac{5}{6}$ of 24.

7. 60 is $\frac{1}{7}$ less than what number?

8. 48 is $\frac{1}{5}$ more than what number?

9. How much will $\frac{2}{3}$ of a ton of coal cost at $4.50 a ton?

10. If $\frac{1}{3}$ of a dozen of eggs costs 6 cents, how much will $2\frac{2}{3}$ doz. cost?

11. How much will $\frac{3}{4}$ of a dozen of eggs cost at $\frac{2}{3}$ of a cent each?

12. If $\frac{1}{6}$ of my yearly salary is $800, how much do I earn in 2 months?

13. If I should have 20 oranges, and should sell $\frac{1}{4}$ of them and $\frac{2}{5}$ of them, how many should I have left?

14. After selling $\frac{2}{3}$ and $\frac{1}{6}$ of my oranges I had 5 oranges left. How many had I at first?

15. If 40 sq. rd. of land cost $20, how much will $2\frac{1}{4}$ acres cost?

16. If an acre and a half of land cost $30, how much will $2\frac{1}{2}$ acres cost?

17. If $2\frac{1}{2}$ quarts of milk cost 15 cents, how much will $2\frac{1}{2}$ gallons cost?

18. Find the area of a rectangle which is $5\frac{1}{2}$ yd. long and $6\frac{1}{2}$ ft. wide.

Construction

Review pages 78, 85, 92, 117, 122, 123, and 151.

1. Construct a right triangle with an angle of 45°. What should be the size of the third angle? Measure it.

2. Construct a right triangle with an angle of 35°.

3. Construct a triangle with an angle of 32° and another angle of 67°.

4. Construct a triangle with an obtuse angle and an angle of 25°.

5. Construct a parallelogram with an angle of 85° and an angle of 95°.

6. Construct a parallelogram with an angle of 70°.

7. Construct an isosceles triangle with an angle of 40°.

8. Construct an isosceles triangle with an angle of 50° at the vertex.

9. Construct an isosceles triangle with a side $2\frac{1}{2}$ inches long.

10. Construct an equilateral triangle.

11. Construct a 3-inch square. Find the middle points of all the sides. Draw dotted lines connecting the middle points of the opposite sides. From the point of intersection of these two lines measure the distance to the middle points of the sides. Inscribe a circle within the square. Find the circumference of the circle. Find the area of the circle. Find the difference between the area of the square and the area of the circle.

12. Construct a square $3\frac{1}{2}$ inches long. Inscribe a circle in the square. Find the circumference of the circle. Find the area of the circle.

155

Fractions — Drill Work

See pages 18, 38, and 120.

Divide directly by inspection:

1. $\frac{5}{8} \div 5$.
2. $\frac{14}{15} \div 7$.
3. $12 \div \frac{1}{2}$.
4. $15 \div \frac{1}{4}$.
5. $\frac{7}{10} \div \frac{1}{10}$.
6. $\frac{15}{16} \div \frac{3}{16}$.
7. $\frac{24}{27} \div \frac{4}{27}$.

8. $16 \div \frac{1}{5}$.
9. $48 \div \frac{1}{3}$.
10. $\frac{36}{37} \div 9$.
11. $\frac{49}{50} \div \frac{7}{50}$.
12. $6\frac{1}{2} \div \frac{1}{2}$.
13. $9\frac{3}{8} \div \frac{1}{8}$.
14. $\frac{28}{35} \div \frac{4}{35}$.

15. $1\frac{1}{2} \div \frac{3}{4}$.
16. $6 \div \frac{3}{4}$.
17. $4 \div 1\frac{1}{3}$.
18. $10 \div 2\frac{1}{2}$.
19. $12\frac{1}{2} \div 2\frac{1}{2}$.
20. $20 \div 3\frac{1}{3}$.
21. $18 \div 1\frac{1}{2}$.

Change to improper similar fractions and divide:

22. $7\frac{1}{3} \div 2\frac{1}{2}$.
23. $12\frac{3}{4} \div 5$.
24. $17\frac{5}{8} \div 7$.
25. $16\frac{1}{7} \div 2\frac{1}{2}$.
26. $20 \div 5\frac{2}{5}$.
27. $25 \div 6\frac{1}{7}$.
28. $28\frac{9}{10} \div 4\frac{2}{5}$.

29. $17\frac{2}{3} \div 4\frac{2}{3}$.
30. $22\frac{5}{9} \div 3\frac{1}{6}$.
31. $34\frac{1}{3} \div 5\frac{2}{7}$.
32. $48\frac{3}{11} \div 7\frac{1}{2}$.
33. $56\frac{9}{10} \div 1\frac{1}{3}$.
34. $60\frac{7}{12} \div 13\frac{1}{2}$.
35. $64\frac{13}{14} \div 2\frac{1}{4}$.

36. $100 \div 3\frac{1}{2}$.
37. $125 \div 7\frac{5}{8}$.
38. $128\frac{2}{3} \div \frac{5}{8}$.
39. $208\frac{9}{10} \div 12$.
40. $250 \div \frac{15}{16}$.
41. $186\frac{8}{9} \div 100$.
42. $312\frac{1}{2} \div 112\frac{1}{2}$.

Divide by the more convenient method:

43. $\frac{15}{21} \div 5$.
44. $\frac{24}{28} \div 7$.
45. $14 \div \frac{1}{2}$.
46. $14 \div \frac{5}{8}$.
47. $6\frac{1}{8} \div \frac{1}{3}$.
48. $6\frac{1}{8} \div \frac{1}{4}$.
49. $12\frac{9}{10} \div \frac{7}{10}$.

50. $37\frac{1}{2} \div 2\frac{1}{2}$.
51. $45\frac{5}{8} \div 3\frac{2}{3}$.
52. $50 \div 2\frac{1}{2}$.
53. $56\frac{1}{4} \div 4\frac{1}{2}$.
54. $16 \div \frac{1}{3}$.
55. $16\frac{1}{3} \div 3$.
56. $37 \div \frac{1}{8}$.

57. $100 \div 2\frac{1}{2}$.
58. $30 \div 3\frac{1}{3}$.
59. $100 \div 2\frac{2}{3}$.
60. $17\frac{1}{4} \div 100$.
61. $\frac{3}{4} \div 7$.
62. $\frac{28}{30} \div 10$.
63. $150 \div 3\frac{7}{10}$.

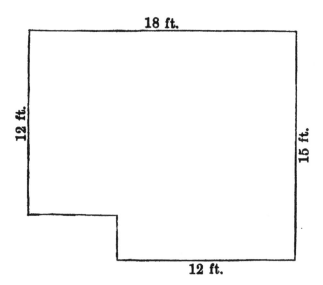

The diagram represents the floor plan of a room which is 9 feet high.

1. What is the distance around the room? ʃ ˥

2. If no account is made of doors, windows, etc., how many strips of paper 18 inches wide will it take to go across the north side?

3. How many strips will it take to go entirely around the room?

4. If the rolls of paper contain 8 yards each, how many rolls will it take to paper the room?

5. If the doors, windows, etc., make a difference of 18 yards of paper, how many rolls will it take for the room?

6. How many yards of border paper will it take to go around the top of the room?

7. How many square feet of plastering are there in the ceiling of the room?

8. What will be the cost of plastering the walls and the ceiling at $12\frac{1}{2}$ cents a square yard, if no allowance is made for doors, windows, etc.?

157

Factors

1. Find the factors of 60.   Find its smallest factors.

$$4\overline{)60} \quad 60 = 4 \times 15 \qquad \begin{array}{l} 2\overline{)60} \\ 2\overline{)30} \\ 3\overline{)15} \\ \quad 5 \end{array} \quad 60 = 2 \times 2 \times 3 \times 5$$

4 and 15 are factors of 60, but 2, 2, 3, and 5 are the *smallest factors* of 60.

To find the smallest factors of a number divide the number and the quotients successively by small numbers, 2's, 3's, 5's, 7's, etc., as far as possible. The several divisors, together with the last quotient, are the factors required.

A number may be divided by another by removing all the factors of the latter from the factors of the former.

The factors of 60 are 2, 2, 3, and 5.   The factors of 15 are 3 and 5. The quotient of 60 divided by 15 is the product of the factors remaining after the factors of 15 are removed from the factors of 60.

Find the smallest factors of:

2. 42.	**6.** 108.	**10.** 972.
3. 72.	**7.** 240.	**11.** 480.
4. 216.	**8.** 300.	**12.** 396.
5. 432.	**9.** 324.	**13.** 468.

14. Divide the product of 2, 3, 3, 5, and 7 by the product of 2, 3, and 5.

15. Factor 216 and 18, and divide by removing factors.

16. Divide 540 by 27 by removing factors.

17. Divide 1728 by 36 by removing factors.

18. Divide 462 by 42 by removing factors.

19. Divide 6912 by 72 by removing factors.

158

Miscellaneous Problems

1. There are 6 ft. in a fathom. How many fathoms are there in a mile?

2. The average depth of the Atlantic Ocean is from 1500 to 3000 fathoms. If we consider the general average to be 2200 fathoms, about how many miles deep is the Atlantic?

3. How long would it take for the weight of a sounding line, sinking at the rate of 10 ft. a second, to reach a depth of 2000 fathoms?

4. The greatest depth yet found in the ocean is at a point near Guam in the north Pacific. At this point the ocean is 5269 fathoms deep. How many miles deep is this?

5. Mt. Everest, the highest mountain in the world, is 29,002 ft. high. How does the greatest depth of the ocean compare with the height of the highest mountain?

6. Sea water weighs about 1.02 times as much as pure water. What is the weight of a cubic foot of sea water?

7. How many ounces does the water of the ocean press downward upon a square inch at the depth of 1 foot?

Divide the weight of a cubic foot of sea water by the number of square inches in a square foot.

8. Find how many pounds the water presses upon a square inch in the ocean at a depth of 100 ft.

9. What is the pressure upon 1 sq. ft. at a depth of 250 ft.?

10. The deepest sounding yet made in the Atlantic Ocean is 4561 fathoms. The ocean has this depth at a point 100 miles north of Porto Rico. Find what number of pounds water presses upon 1 sq. in. at that depth.

Wood Measure

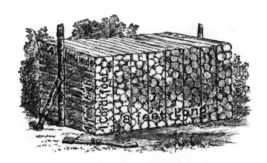

When wood is sent to market it is usually cut into sticks 4 ft. long. A pile of such wood which is 8 ft. long and 4 ft. high contains a **cord**.

1. How many square feet of surface are there on one side of a pile which contains a cord?

2. How many square feet of surface are there on one side of a pile which is 56 ft. long and 8 ft. high?

3. How many cords are there in this pile?

4. How many cords are there in a pile of wood one side of which is 16 ft. long and 6 ft. high?

If the wood is cut 4 ft. long and piled 4 ft. high, a section of the pile 1 ft. long is called a cord foot.

5. How many cord feet are there in a cord of wood?

6. How many cubic feet make a cord foot?

7. How many cord feet are there in a pile 12 ft. long, 8 ft. wide, and 4 ft. high?

8. How many cubic feet are there in a cord of wood?

9. How many cubic feet are there in $5\frac{3}{4}$ cords?

10. What part of a cord is 32 cubic feet?

11. How many cords are there in 320 cubic feet?

12. Find how many cords there are in a pile which is 40 ft. long, 6 ft. high, and 4 ft. wide.

13. How many cords in a pile 28 ft. × 6 ft. × 5 ft.?

160

Decimal Fractions

$$\tfrac{4}{10} = \tfrac{40}{100} = \tfrac{400}{1000}. \qquad\qquad .4 = .40 = .400.$$

Since ciphers annexed to a decimal fraction do not change its value, they may be annexed at any time for the sake of convenience.

If the number of decimal figures in the dividend is not equal to the number in the divisor, it may be made equal by adding ciphers. When there is a remainder in division, and it is desirable, for the sake of greater accuracy, to extend the number of figures in the decimal fraction of the quotient, ciphers may be annexed to the dividend indefinitely.

1. Divide 375 by .125.

The divisor is thousandths. In the dividend there are 375,000 thousandths. The quotient is 3000.

$$\begin{array}{r} 3000 \\ \hline .125)\overline{375.000} \end{array}$$

Always place the decimal point in the quotient immediately after that figure of the dividend whose denomination is the same as that of the right hand figure of the divisor.

2. Divide 4.7 by 3.

Having placed the decimal point at the proper place in the quotient, we may continue the division indefinitely by imagining ciphers placed after the dividend.

$$\begin{array}{r} 3)4.7 \\ \hline 1.5666 + \end{array}$$

One third of 4 is 1 with 1 remaining. One third of 17 tenths is 5 tenths with 2 tenths or 20 hundredths remaining. One third of 20 hundredths is 6 hundredths, with 2 hundredths remaining, etc.

Divide :

3. 25 by .5.

4. 34 by .002.

5. 57.5 by .25.

6. .16 by .004.

7. 1562.5 by .625.

8. 45 by 900.

9. .17 by 8.

10. 5.29 by 3.

11. .0169 by .013.

12. 8 by .0004.

13. $9.45 \times 3.6 = ?$

14. $.645 \times .2103 = ?$

15. $25.874 \times 100 = ?$

16. $5.45 \times 7\tfrac{2}{5} = ?$

17. $24.564 \times 143\tfrac{3}{4} = ?$

18. $9 \div .13 = ?$

Miscellaneous Problems

1. If 2 barrels of apples cost \$$4\frac{4}{5}$, how much will $10\frac{1}{2}$ barrels cost?

2. If $\frac{3}{4}$ of a barrel of sugar costs \$$9\frac{3}{8}$, what is the cost of a barrel?

3. If $\frac{2}{3}$ of a barrel of sugar costs \$$8\frac{4}{5}$, how much will $2\frac{1}{2}$ barrels cost?

4. A man lost $\frac{3}{8}$ of his money, and afterwards found $\frac{2}{3}$ of what he had lost. What part of all his money did he find?

5. What part of all his money had he after finding $\frac{2}{3}$ of what he had lost?

6. If the part which he did not find was \$4, how much money had he at first?

7. A certain farm contains 80 acres. $\frac{1}{5}$ of it is used for corn, $\frac{2}{3}$ for wheat, and the remainder for grass. How many acres are used for grass?

8. $\frac{1}{3}$ of a certain farm is used for oats, $\frac{3}{8}$ for grazing, and the remainder for corn. What part of the farm is used for corn?

9. If 56 acres are used for corn, how many acres are there in the farm?

10. $\frac{3}{5}$ of a certain farm is used for corn, and $\frac{1}{2}$ of the remainder for wheat. If 20 acres are used for wheat, how many acres are used for corn?

11. A merchant sold goods for $\frac{5}{8}$ of their cost, and received \$25. What was the cost?

12. A merchant sold wheat for $\frac{5}{4}$ of the cost, and received \$30. What was the cost?

13. A dealer sold hay at $\frac{4}{5}$ of the cost, and lost \$4 a ton. What was the cost per ton?

Mental Problems

1. If a man earns $2½ a day and spends $10½ a week, in how many weeks will he save $45?

2. A man earns $2 a day besides his board when he works, and when idle pays 50 cents a day for his board. At the end of 6 days he has $7. How many days was he idle?

3. 2 men start at the same time to travel in the same direction. One travels 3½ miles an hour, and the other 4¼ miles an hour. How far apart will they be at the end of 10 hours?

4. A man traveled 8 hours at the rate of 5¼ miles an hour, and then returned to the point from which he started at the rate of 3½ miles an hour. How long was it from the time of starting until his return?

5. A man walked a distance of 44 miles. He started at 7 A.M. and reached his destination at 4 P.M. He stopped twice on the way to rest for a half hour. At what rate per hour did he walk?

6. On a river which flows at the rate of 2 miles an hour a man who can row 6 miles an hour in still water rows down the river for 5 hours. How far is he then from the starting point?

In rowing down, the current adds 2 miles an hour to his rate of speed; in rowing back it retards his speed 2 miles an hour.

7. How long will it take him to row back again?

8. If a man can row down stream 8 miles an hour, and up stream only 2 miles an hour, how fast does the water flow?

The difference between the two is twice the rate of the current. Why?

9. If a man can row down stream 7 miles an hour, and up stream 3 miles an hour, how many miles an hour can he row in still water?

163

Ratio

Review page 111.

1. What is the ratio of 75 to 5?
2. What is the ratio of 20 to 40?
3. What is the ratio of 9 weeks to 3 days?
4. What is the ratio of 4 rods to 8 inches?
5. What is the ratio of 96 lb. to 4 lb. 6 oz.?

Find the ratio of:

6. $\frac{1}{2}$ to $\frac{1}{4}$
7. $\frac{3}{8}$ to $\frac{3}{4}$
8. $\frac{9}{10}$ to $\frac{1}{2}$
9. $\frac{14}{16}$ to $\frac{2}{8}$
10. $\frac{17}{18}$ to $\frac{1}{6}$

11. $3\frac{4}{5}$ to $\frac{1}{5}$
12. $48\frac{4}{8}$ to 4
13. $125\frac{15}{16}$ to 5
14. 25 to $\frac{1}{4}$
15. 14 to $\frac{2}{8}$

Find the ratio of:

16. A 4-inch square to a square foot.
17. A square yard to a square foot.
18. A rectangle 15 in. × 10 in. to a 5-inch square.
19. A rectangle 8 in. × 20 in. to a rectangle 6 ft. × 4 ft.
20. A foot cube to a 3-inch cube.
21. A 9-inch cube to a cube 4 ft. 3 in. long.

Find the ratio of:

22. The area of a circle 10 inches in diameter to one 2 feet in diameter.
23. The number of feet of lumber in a board 20 ft. long, 6 in. wide to that in a board 10 ft. long, 3 in. wide.
24. The number of cords of wood in a pile 32 ft. × 12 ft. × 6 ft. to that in a pile 8 ft. × 8 ft. × 4 ft.

164

Fractions — Addition and Subtraction

1. How many thirds are there in one fifteenth?

2. How many thirtieths in one fifth?

3. Express three tenths as thirtieths.

4. How much is one fifth less one fifteenth?

5. How much is one fifteenth and one thirtieth?

6. How much is one sixth less one thirtieth?

7. What is the sum of one fifth and one sixth?

Make the fractions similar:

Determine the size of the terms by inspection, and do the work mentally as far as possible.

8. $\frac{1}{2}, \frac{8}{10}, \frac{1}{15}, \frac{7}{30}$ 11. $\frac{1}{3}, \frac{5}{6}, \frac{4}{9}, \frac{1}{2}$ 14. $\frac{4}{5}, \frac{5}{6}, \frac{7}{15}$

9. $\frac{2}{3}, \frac{2}{9}, \frac{7}{18}, \frac{5}{36}$ 12. $\frac{1}{8}, \frac{2}{5}, \frac{3}{20}, \frac{9}{40}$ 15. $\frac{1}{5}, \frac{3}{8}, \frac{7}{20}$

10. $\frac{1}{7}, \frac{1}{6}, \frac{1}{3}, \frac{1}{42}$ 13. $\frac{1}{8}, \frac{5}{6}, \frac{3}{8}, \frac{1}{16}$ 16. $\frac{1}{2}, \frac{3}{10}, \frac{2}{25}$

Add:

17. $\frac{1}{2}, \frac{4}{5}, \frac{7}{10}$ 23. $7\frac{1}{2}, 8\frac{3}{5}$ 29. $\frac{5}{6}, \frac{1}{16}, \frac{7}{48}$

18. $\frac{1}{3}, \frac{5}{6}, \frac{4}{9}$ 24. $9\frac{5}{6}, 4\frac{2}{9}$ 30. $\frac{1}{2}, \frac{5}{12}, \frac{7}{16}$

19. $\frac{1}{5}, \frac{5}{6}, \frac{11}{30}$ 25. $12\frac{3}{8}, 14\frac{1}{5}$ 31. $\frac{2}{3}, \frac{3}{8}, \frac{1}{24}$

20. $\frac{1}{2}, \frac{5}{6}, \frac{2}{15}$ 26. $\frac{2}{5}, \frac{1}{25}, \frac{8}{50}$ 32. $10\frac{3}{5}, 9\frac{7}{50}$

21. $\frac{1}{8}, \frac{3}{5}, \frac{8}{40}$ 27. $\frac{7}{10}, \frac{1}{2}, \frac{4}{25}$ 33. $13\frac{1}{3}, 8\frac{7}{16}$

22. $\frac{5}{8}, \frac{1}{5}, \frac{8}{10}$ 28. $\frac{1}{2}, \frac{4}{5}, \frac{9}{25}$ 34. $15\frac{5}{8}, 12\frac{1}{6}$

Subtract:

35. $\frac{1}{2}$ from $\frac{8}{6}$ 38. $\frac{1}{8}$ from $\frac{1}{5}$ 41. $5\frac{1}{2}$ from $8\frac{4}{5}$

36. $\frac{1}{6}$ from $\frac{1}{6}$ 39. $\frac{3}{8}$ from $\frac{7}{10}$ 42. $7\frac{1}{40}$ from $12\frac{3}{8}$

37. $\frac{4}{5}$ from $\frac{5}{6}$ 40. $\frac{4}{5}$ from $\frac{23}{24}$ 43. $10\frac{1}{4}$ from $15\frac{3}{50}$

165

Hundredths

1. How many hundredths in one fiftieth?

2. How many hundredths in three twenty-fifths?

3. How many hundredths in two wholes?

4. How many hundredths in forty-five wholes?

5. How much is $\frac{3}{50}$ and $\frac{1}{100}$?

6. How much is $\frac{6}{25}$ and $\frac{24}{100}$?

7. How much is $\frac{9}{50}$ less $\frac{8}{100}$?

8. Add $\frac{3}{10}$, $\frac{1}{25}$, $\frac{11}{100}$.

9. From $\frac{99}{100}$ take $\frac{18}{20}$.

10. Express as hundredths: $\frac{3}{50}$; $\frac{7}{25}$; $\frac{9}{20}$; $\frac{3}{10}$; $\frac{4}{5}$; $\frac{3}{4}$.

11. Express as hundredths: $\frac{1}{2}$; $\frac{1}{4}$; $\frac{1}{5}$; $\frac{9}{10}$; $\frac{3}{5}$; $\frac{19}{20}$.

12. What is $\frac{1}{5}$ of 20? $\frac{3}{5}$ of 20? $\frac{5}{8}$ of 40? $\frac{1}{25}$ of 50?

13. What is $\frac{4}{25}$ of 50? $\frac{1}{50}$ of 50? $\frac{13}{50}$ of 50? $\frac{21}{50}$ of 100?

14. What is $\frac{1}{100}$ of 100? $\frac{12}{100}$ of 100? $\frac{1}{100}$ of 300? $\frac{7}{100}$ of 300?

15. What is $\frac{42}{100}$ of 400? $\frac{17}{100}$ of \$5? $\frac{25}{100}$ of \$30. $\frac{75}{100}$ of \$80?

16. A farmer, having 200 sheep, lost $\frac{37}{100}$ of them. How many did he lose?

17. A farmer, having 300 sheep, lost $\frac{65}{100}$ of them. How many had he left?

18. If $\frac{6}{100}$ of the principal is paid each year for the use of money, how much is paid for the use of \$250?

19. If $\frac{1}{32\frac{1}{2}}$ of the principal is paid as interest, how much interest is paid on a principal of \$1000?

20. If I receive as interest every 6 months $\frac{1\frac{1}{2}}{100}$ of what money I have in the savings bank, and have \$500 in the bank, how much interest do I receive each year?

166

Miscellaneous Problems

1. What is the area of a rectangle which is 9 ft. 6 in. long and 6 ft. 4 in. wide?

2. What is the area of a parallelogram if one side is $16\frac{4}{5}$ in. long and the perpendicular distance from this side to the opposite side is 10 in.?

3. What is the area of a triangle whose base is 7.5 yd. and altitude 8.5 ft.?

4. The roof of a barn is 15 ft. higher than the tops of the posts. The barn is 32 ft. wide. What is the area of the gables?

5. How many feet of boards will it take to board one end of a barn, including the gable, if the barn is 34 ft. wide and 18 ft. high to the eaves, and the roof extends 16 ft. above the eaves?

6. Find the circumference of a circle whose diameter is 9 ft.

7. Find the diameter of a circle whose circumference is 100 ft.

8. It is 50 miles around a circular lake. How long will it take to row across it at the rate of 3 miles an hour?

9. Find the area of a circle whose radius is 5 yards.

10. Find the area of the largest circle which can be inscribed in a square 12 ft. long.

11. What would be the area of the part outside the circle and within the square?

12. What is $\frac{5}{8}$ of 400? $\frac{7}{12}$ of 600? $\frac{48}{60}$ of 1000?

13. What is $\frac{1}{100}$ of 200? $\frac{9}{100}$ of 200? $\frac{17}{100}$ of 300?

14. Find $\frac{87}{100}$ of 2500. $\frac{48}{100}$ of 3800.

15. Find $\frac{5\frac{1}{2}}{100}$ of 800. $\frac{6\frac{1}{4}}{100}$ of 650.

167

Per Cent

It is customary for many purposes to express fractional parts, when possible, as hundredth parts. The fraction is then called per cent, and may be indicated by a special symbol.

The fraction seven hundredths may be written $\frac{7}{100}$, .07, 7 per cent, or 7%.

1. What is three hundredths of 200?

First find one hundredth.

2. What is sixty-five hundredths of 1000?

3. Find fifteen hundredths of 2300.

4. Find seventy-seven hundredths of 4500.

5. What is $\frac{1}{100}$ of 200? $\frac{1}{100}$ of 1200? $\frac{5}{100}$ of 2400?

6. What is $\frac{3}{100}$ of 350? $\frac{12}{100}$ of 500? $\frac{3}{100}$ of 600?

7. What is 1% of 200?   3% of 200?   7% of 500?

8. What is 7% of 300?   9% of 400?   25% of 1000?

9. What is $\frac{1}{100}$ of 400? $\frac{3}{100}$ of 400?   8% of 4000?

10. If shoes which cost $3 a pair are sold at a gain of 10%, how much is gained on each pair?

11. If umbrellas are marked to sell at $2, and are afterwards marked down 5%, how much is the price reduced?

12. What is the selling price of an article which cost $6 and is sold at a gain of 20%?

13. If I buy oranges for $4 a box and sell them so as to gain 50%, how much shall I make on each box?

14. If there are 500 pupils belonging to a school and 5% of them are absent, how many are absent?

15. If there are 20% more girls in school than boys, and there are 250 boys, how many girls are there?

168

Per Cent

1. Find 6% of 50.

1 per cent of 50 is ½. 6 per cent of 50 is 6 times ½, or 3.

What is :

2. 1 per cent of 300 ?

3. 5 per cent of 300 ?

4. 8 per cent of 200 ?

5. 10 per cent of 500 ?

6. 17 per cent of 800 ?

7. 23% of 2500 ?

8. 32% of 3100 ?

9. 21½% of 4400 ?

10. 43¼% of 5600 ?

11. 51% of 6500 ?

12. Find the number of which 6 is 2 per cent.

Since 6 is 2% of the number, 1% of the number is ½ of 6, or 3;
100% of the number or the whole number is 100 × 3, or 300.

Find the number of which :

13. 8 is 2 per cent.

14. 15 is 5 per cent.

15. 60 is 4 per cent.

16. 84 is 7 per cent.

17. 72 is 9 per cent.

18. 140 is 7%.

19. 700 is 35%.

20. 120 is 30%.

21. 450 is 25%.

22. 480 is 16%.

What is :

23. 1 per cent of 50 ?

24. 6 per cent of 50 ?

25. 80 per cent of 50 ?

26. 1 per cent of 25 ?

27. 65% of 50 ?

28. 84% of 50 ?

29. 38% of 25 ?

30. 72% of 25 ?

Find the number of which.:

31. 50 is 1 per cent.

32. 60 is ½ per cent.

33. 40 is ⅛ per cent.

34. 22 is ¼ per cent.

35. 1200 is 60%.

36. 2870 is 35%.

37. 5625 is 25%.

38. 5665 is 55%.

169

Review Problems

1. How many degrees are there in a semicircle?

2. How many right angles does it take to make a semicircle?

3. If an angle of $22\frac{1}{2}°$ is taken from a right angle, how many degrees will remain?

4. In a right triangle one of the acute angles is $42\frac{1}{2}°$. How large is the other acute angle?

5. Two angles of a triangle are $103\frac{1}{2}°$ and $16\frac{1}{2}°$. How many degrees are there in the third angle?

6. How many degrees are there in each of the angles of an equilateral triangle?

7. The angle at the vertex of an isosceles triangle is $48\frac{1}{4}°$. How large is each angle at the base?

8. What is $\frac{4}{100}$ of 900? .25 of 340?

9. What is 50 per cent of 61,200? 5% of $200?

10. A man's salary is $2000, and he saves 15% of it. How much does he save?

11. In an orchard containing 500 trees 30% of the trees are apple trees, and the remainder pear trees. How many pear trees are there?

12. A farmer having 150 chickens lost 18% of them. How many had he left?

13. In a school containing 350 pupils 4% are absent. How many are present?

14. If 60% of the ore from a certain mine is lead, how much lead is there in a ton of the ore?

15. If a merchant should buy flour at $6.00 a barrel, and sell it so as to gain 15%, how much would he gain on 50 barrels?

Fractions — Drill Work

Add:

1. $\frac{3}{4} + \frac{2}{3} + \frac{5}{6}$ 6. $\frac{1}{2} + \frac{1}{10} + \frac{1}{15}$

2. $\frac{2}{3} + \frac{1}{15} + \frac{4}{5}$ 7. $\frac{1}{4} + \frac{2}{9} + \frac{5}{18}$

3. $\frac{5}{9} + \frac{1}{3} + \frac{5}{6}$ 8. $\frac{3}{5} + \frac{5}{8} + \frac{1}{20}$

4. $\frac{9}{10} + \frac{1}{2} + \frac{4}{15}$ 9. $\frac{8}{32} + \frac{5}{8} + \frac{3}{4}$

5. $\frac{1}{3} + \frac{1}{6} + \frac{1}{18}$ 10. $\frac{1}{14} + \frac{5}{7} + \frac{1}{2}$

Subtract:

11. $\frac{5}{6} - \frac{1}{5}$ 16. $\frac{5}{9} - \frac{1}{2}$

12. $\frac{2}{3} - \frac{2}{7}$ 17. $\frac{6}{7} - \frac{1}{2}$

13. $\frac{3}{4} - \frac{5}{8}$ 18. $\frac{5}{7} - \frac{2}{8}$

14. $\frac{11}{12} - \frac{3}{8}$ 19. $\frac{4}{5} - \frac{3}{7}$

15. $\frac{19}{20} - \frac{4}{5}$ 20. $\frac{2}{3} - \frac{5}{16}$

Multiply:

21. $\frac{6}{7} \times \frac{2}{3}$ 26. $236\frac{5}{8} \times 17$

22. $\frac{9}{10} \times \frac{1}{3}$ 27. $312\frac{4}{5} \times 22$

23. $\frac{11}{12} \times \frac{5}{11}$ 28. $621\frac{1}{7} \times 43$

24. $\frac{9}{10} \times \frac{20}{21}$ 29. $218\frac{3}{4} \times 26$

25. $\frac{6}{7} \times \frac{14}{31}$ 30. $921\frac{5}{9} \times 34$

Divide:

31. $\frac{9}{10} \div 3$ 36. $\frac{7}{10} \div \frac{2}{5}$

32. $\frac{12}{13} \div 4$ 37. $\frac{7}{8} \div \frac{1}{2}$

33. $\frac{15}{31} \div 5$ 38. $\frac{11}{12} \div \frac{2}{3}$

34. $\frac{9}{10} \div \frac{3}{10}$ 39. $\frac{4}{5} \div \frac{1}{3}$

35. $\frac{8}{9} \div \frac{2}{9}$ 40. $\frac{8}{9} \div \frac{1}{4}$

Original Problems

Make problems and solve them:

1. A room 15 ft. long, 12 ft. wide, and 9 ft. high is to be papered.

2. The boards for the floor of the room cost $35 a thousand feet.

3. At some places the Atlantic Ocean is 2000 fathoms deep.

4. Sea water weighs about 1.02 times as much as pure water.

5. A pile of wood is 12 feet long and 6 feet high.

6. A dealer bought a pile of wood 24 feet long, 4 feet wide, and 8 feet high.

7. A car load of wood was 32 ft. long, 10 ft. high, and 8 ft. wide.

8. $\frac{3}{4}$ of a barrel of flour costs $3\frac{3}{4}$.

9. A boy lost $\frac{3}{5}$ of his marbles and afterward found $\frac{2}{3}$ of what he had lost.

10. A dealer sold flour for $\frac{1}{5}$ more than the cost.

11. A dealer sold goods for $\frac{1}{6}$ less than the cost.

12. A man earns $1.50 a day, besides his board, when he works, and when idle pays 50 cents a day for his board.

13. A man traveled 18 miles at the rate of $4\frac{1}{2}$ miles an hour and returned at the rate of 3 miles an hour.

14. A man can row down stream 5 miles an hour and up stream 1 mile an hour.

15. A boy has $\frac{1}{2}$ as much money as his sister, and together they have 24 cents.

16. The difference between $\frac{1}{3}$ and $\frac{1}{4}$ of my money is $5.

17. There are 25% more boys in school than girls.

172

Fractions — Drill Work

See pages 43 and 120.

Make the fractions similar:

1. $\frac{3}{4}, \frac{2}{3}, \frac{5}{6}, 1\frac{1}{12}.$
2. $\frac{1}{5}, \frac{1}{3}, \frac{4}{5}, \frac{2}{3}.$
3. $\frac{3}{2}, \frac{1}{6}, \frac{7}{9}, \frac{1}{8}.$
4. $\frac{2}{5}, \frac{9}{10}, \frac{3}{4}, \frac{1}{2}.$
5. $\frac{7}{8}, \frac{1}{4}, \frac{11}{12}, \frac{5}{6}.$

6. $\frac{1}{4}, \frac{1}{2}, \frac{5}{8}, \frac{7}{16}, \frac{9}{32}.$
7. $\frac{5}{9}, \frac{5}{4}, \frac{8}{18}, \frac{1}{3}, \frac{5}{6}.$
8. $\frac{4}{5}, \frac{3}{8}, \frac{1}{10}, \frac{3}{5}, \frac{19}{20}.$
9. $\frac{1}{2}, \frac{3}{4}, \frac{6}{7}, \frac{3}{14}, \frac{8}{7}.$
10. $\frac{4}{5}, \frac{1}{2}, \frac{2}{3}, \frac{1}{6}, \frac{3}{5}.$

Make similar and add:

11. $\frac{2}{3} + \frac{5}{12} + \frac{3}{4} + \frac{1}{2}.$
12. $\frac{1}{8} + \frac{5}{9} + \frac{1}{6} + \frac{1}{2}.$
13. $\frac{9}{10} + \frac{1}{2} + \frac{4}{5} + \frac{3}{4}.$
14. $\frac{1}{4} + \frac{5}{12} + \frac{1}{6} + \frac{3}{8}.$
15. $\frac{1}{8} + \frac{1}{16} + \frac{1}{4} + \frac{1}{2}.$

16. $3\frac{1}{2} + 5\frac{1}{8} + \frac{5}{9} + \frac{1}{18}.$
17. $\frac{4}{5} + 6\frac{1}{3} + \frac{1}{2} + 2\frac{1}{15}.$
18. $9\frac{1}{4} + 3\frac{1}{2} + \frac{4}{5} + 1\frac{7}{10}.$
19. $4\frac{1}{2} + 11\frac{3}{4} + \frac{1}{14} + \frac{5}{7}.$
20. $17 + 14\frac{1}{5} + 1\frac{1}{6} + \frac{3}{2}.$

Make similar and subtract:

21. $\frac{11}{12} - \frac{3}{4}.$
22. $\frac{14}{15} - \frac{2}{3}.$
23. $\frac{7}{8} - \frac{2}{3}.$
24. $\frac{3}{5} - \frac{1}{4}.$
25. $\frac{5}{6} - \frac{3}{4}.$

26. $9\frac{2}{3} - \frac{1}{5}.$
27. $9\frac{2}{3} - \frac{4}{5}.$
28. $12\frac{2}{3} - 4\frac{1}{6}.$
29. $12\frac{2}{3} - 2\frac{8}{9}.$
30. $10 - 3\frac{2}{5}.$

31. $\frac{9}{10} - \frac{3}{4}.$
32. $\frac{5}{6} - \frac{1}{6}.$
33. $\frac{7}{8} - \frac{4}{5}.$
34. $15\frac{1}{4} - \frac{1}{8}.$
35. $32\frac{1}{7} - 12\frac{1}{4}.$

Make similar and divide:

36. $\frac{11}{12} \div \frac{2}{3}.$
37. $\frac{17}{18} \div \frac{1}{2}.$
38. $\frac{2}{3} \div \frac{1}{4}.$
39. $\frac{4}{5} \div \frac{3}{4}.$
40. $\frac{5}{6} \div \frac{5}{9}.$

41. $7\frac{1}{3} \div 1\frac{1}{5}.$
42. $16\frac{3}{8} \div 4\frac{1}{8}.$
43. $\frac{19}{20} \div 2\frac{4}{5}.$
44. $7 \div 13\frac{5}{7}.$
45. $\frac{29}{30} \div 5.$

46. $\frac{23}{28} \div \frac{2}{7}.$
47. $\frac{1}{2} \div \frac{17}{80}.$
48. $14 \div \frac{8}{11}.$
49. $\frac{27}{40} \div 13.$
50. $17\frac{3}{4} \div 20\frac{4}{5}.$

173

Factors

Review page 158.

1. What are factors of 4? 6? 15? 21?

2. What are factors of 20? 24? 25? 36?

3. Give the smallest factors of 8. 12. 16. 18.

4. Give the smallest factors of 40. 48. 54. 100.

5. If the product of two numbers is 6 and one of the numbers is 2, what is the other number?

6. If the product of two numbers is 35 and one of the numbers is 5, what is the other number?

7. What is one of the two equal factors of 36? 64? 100?

8. What is one of the three equal factors of 27? 64? 125?

9. If two of the factors of a number are 2 and 3, and the number is 18, what is the other factor?

10. If the product of three numbers is 45, and the product of two of the numbers is 9, what is the other number?

11. The area of a square is 25 sq. in. How long is it?

12. The area of a rectangle is 63 sq. ft. and it is 7 ft. long. How wide is it?

13. A rectangular field which is 8 rd. wide contains 80 sq. rd. How long is it?

14. What are the dimensions of a rectangle containing 8 sq. in. if its length is twice its width?

Think of the rectangle as two squares.

15. An oblong is three times as long as it is wide, and contains 48 sq. in. How long is it?

16. If a cube contains 27 cubic inches, how long is it?

174

United States Money

Since 10 cents is one tenth of a dollar and 1 cent one hundredth of a dollar, a number indicating dollars and cents may be regarded as a decimal number.

One tenth of a cent is called a mill and is written in the third place from the decimal point.

$.003 is three mills.
$.015 is one cent five mills, or fifteen mills.
$.234 is twenty-three cents four mills.
$.0043 is four and three tenths mills.
$.00612 is six and twelve hundredths mills.

As there are no coins for mills, they are either omitted in the answer, or if there are five or more mills, they are regarded as one cent. In solving a problem the mills and parts of mills should generally be retained until the answer is obtained.

1. Write: two dollars, eight cents, seven mills; eighteen mills; five and three tenths mills; seven and thirteen hundredths mills.

2. Add: five dollars, four cents, nine mills; three and five tenths mills; four cents, nine and seventeen hundredths mills.

Multiply :

3. $7.569 by 8.

4. $ .097 by 9.

5. $1.085 by 7.

6. $2.438 by 7.8.

7. $.7062 by 5.4.

8. $3.099 by 6.7.

Divide :

9. $4.214 by 7.

10. $.1902 by 6.

11. $7.0184 by 8.

12. $25.134 by 5.3.

13. $8.25 by 3.4.

14. $19.038 by 7.8.

Percentage — Fractions

50%				50%				Halves
25%		25%						Fourths
12½%	12¼%	12½%	12¼%					Eighths
16⅔%		16¼%		16⅔%				Sixths
33⅓%			33⅓%			33⅓%		Thirds

Change to fractions having smaller terms:

1. $\dfrac{10}{100}$ 5. $\dfrac{50}{100}$ 9. $\dfrac{75}{100}$ 13. $\dfrac{62\frac{1}{2}}{100}$

2. $\dfrac{30}{100}$ 6. $\dfrac{12\frac{1}{2}}{100}$ 10. $\dfrac{37\frac{1}{2}}{100}$ 14. $\dfrac{87\frac{1}{2}}{100}$

3. $\dfrac{80}{100}$ 7. $\dfrac{33\frac{1}{3}}{100}$ 11. $\dfrac{60}{100}$ 15. $\dfrac{70}{100}$

4. $\dfrac{25}{100}$ 8. $\dfrac{70}{100}$ 12. $\dfrac{16\frac{2}{3}}{100}$ 16. $\dfrac{66\frac{2}{3}}{100}$

Change to hundredths:

17. $\frac{1}{10}$ 21. $\frac{1}{4}$ 25. $\frac{1}{8}$ 29. $\frac{5}{8}$

18. $\frac{5}{10}$ 22. $\frac{1}{5}$ 26. $\frac{3}{8}$ 30. $\frac{7}{8}$

19. $\frac{8}{10}$ 23. $\frac{3}{4}$ 27. $\frac{1}{3}$ 31. $\frac{5}{6}$

20. $\frac{4}{5}$ 24. $\frac{3}{5}$ 28. $\frac{1}{6}$ 32. $\frac{9}{10}$

33. What per cent of a number is $\frac{1}{8}$ of it? $\frac{3}{8}$? $\frac{5}{8}$?

34. What per cent of a sum of money is $\frac{1}{3}$ of it? $\frac{1}{6}$ of it? $\frac{2}{3}$ of it?

35. What part of a number is 25% of it? 75%? 40%?

36. What part of a sum of money is 20% of it? 16⅔% of it? 87½% of it?

37. What is 16⅔% of 60? 37½% of 72? 62½% of 16?

38. If I sell goods, which cost me $2.40, at a gain of 12½%, how much do I gain?

39. If I sell goods, which cost me $3.60, at a loss of 16⅔%, what is the selling price?

176

Percentage

What is :

1. 50 per cent of 200 ?
2. 50 per cent of 1200 ?
3. 50 per cent of 3240 ?
4. 25 per cent of 4800 ?
5. 12½ per cent of 2480 ?
6. 16⅔ per cent of 5646 ?

7. 33⅓% of 1566 ?
8. 75% of 6644 ?
9. 20% of 9560 ?
10. 66⅔% of 3120 ?
11. 37½% of 8424 ?
12. 62½% of 9256 ?

Find the number of which :

13. 365 is 50 per cent.
14. 472 is 25 per cent.
15. 392 is 12½ per cent.
16. 718 is 16⅔ per cent.
17. 524 is 20 per cent.
18. 390 is 37½ per cent.

19. 1000 is 62½%.
20. 2400 is 66⅔%.
21. 2870 is 87½%.
22. 9366 is 60%.
23. 4880 is 80%.
24. 2898 is 90%.

What is :

25. 10 per cent of 1260 ?
26. 20 per cent of 3645 ?
27. 25 per cent of 1148 ?
28. 30 per cent of 2130 ?
29. 40 per cent of 3875 ?
30. 60 per cent of 4965 ?

31. 100% of 1500 ?
32. 200% of 1275 ?
33. 150% of 4236 ?
34. 250% of 2240 ?
35. 225% of 3448 ?
36. 125% of 2364 ?

Find the number of which :

37. 375 is 10 per cent.
38. 439 is 20 per cent.
39. 586 is 25 per cent.
40. 345 is 30 per cent.
41. 940 is 40 per cent.
42. 645 is 60 per cent.

43. 3840 is 12½%.
44. 2832 is 16⅔%.
45. 5862 is 33⅓%.
46. 7269 is 37½%.
47. 4855 is 62½%.
48. 4494 is 87½%.

Miscellaneous Problems

1. The selling price of goods is $1.15. If I should reduce the price 12½ cents, I should still sell at a profit of 4½ cents. What was the cost?

2. My price for cloth is 83 cents per yard. If I should reduce the price 12 cents a yard, I should lose 4 cents a yard. What was the cost?

3. For how much each must I sell 8 coats that cost $7.25 each, to gain $4 in all?

4. How much shall I make if I buy a box containing 150 oranges for $2.25 and sell the oranges at 24 cents a dozen?

5. A man bought a cow for $50. He paid 25 cents a day for hay and meal to feed her, and sold each day's milk for 45 cents. After keeping the cow 10 days, he sold her for $55. How much did he make by the transaction?

6. A merchant bought 8 barrels of apples at $3¼ a barrel. 3 barrels decayed, and he sold the remainder at $4 a barrel. How much did he lose?

7. In 3 pieces of cloth there are 24⅝ yd., 32⅝ yd., and 19$\frac{7}{12}$ yd. How many yards are there in all?

8. How many gallons of water will a tank contain which is 28 in. long, 15 in. wide, and 16 in. deep?

9. Find the number of miles in 1440 rods.

10. Find the number of rods in 1980 inches.

11. If from a piece of cloth which originally contained 44 yd. there have been sold 8½ yd., 6⅔ yd., 2½ yd., and 10⅛ yd., what will be the cost of the remainder at 12½ cents a yard?

Percentage

1. If I have 50 oranges and sell 10% of them, how many do I sell?

2. A horse which cost $200 was sold at a gain of 25%. How much was the gain? For how much was the horse sold?

3. If goods which cost me $8 are sold for $7, what part of the cost do I lose? What per cent of the cost do I lose?

4. If I sell 50% and 30% of my goods, what per cent shall I have left?

5. If I sell 75% and $12\frac{1}{2}$%, what fractional part shall I have left?

6. What is the difference between $37\frac{1}{2}$% of anything and $12\frac{1}{2}$% of it?

7. Cloth which cost 50 cents a yard was sold at a loss of 15%. For what per cent of the cost was it sold? What was the selling price?

8. What is $87\frac{1}{2}$% of 320? $62\frac{1}{2}$% of 960?

9. What is $\frac{15}{100}$ of 60? .$12\frac{1}{2}$ of 32? $16\frac{2}{3}$% of 72?

10. What is $\frac{66\frac{2}{3}}{100}$ of 45? .23 of 200? 80% of 40?

11. A bill of goods amounting to $25.25 is discounted 20%. How much is the reduction?

12. Goods which were sold at $6.30 are advanced $33\frac{1}{3}$%. What is the selling price now?

13. The amount of rainfall in April was $37\frac{1}{2}$% more than in March. In March it was 4 inches. How much was it in April?

14. There were $16\frac{2}{3}$% less pleasant days in May than in April. There were 18 pleasant days in April. How many were there in May?

179

Percentage

1. What is 20% of 455 lb.?

2. What is 25% of 8 bu. 2 pk. 4 qt.?

3. Find $12\frac{1}{2}$% of 25 lb. 8 oz.

4. Find $33\frac{1}{3}$% of 7 gal. 2 qt.

5. A merchant bought 500 lb. of tea for $200, and sold 60% of it at 48 cents a pound, and the remainder at $52\frac{1}{2}$ cents a pound. How much did he gain?

6. I bought a horse for $300, and sold him at a loss of $16\frac{2}{3}$%. How much did I lose?

7. A wholesale dealer bought 2 car loads of flour containing 120 barrels each. He sold 25% of it to one customer, 20% to another, and 30% to another. How many barrels remained?

8. A bushel of wheat weighs 60 lb. If wheat shrinks 5% in drying after it has been threshed, what will be the weight of 500 bushels after it becomes dry?

9. A piece of real estate which was formerly worth $5800 has depreciated 7% in value. What is its present value?

10. If a city containing 20,352 people increases each year $12\frac{1}{2}$%, what is the increase the first year?

11. If gloves which were formerly sold at $1.50 have been marked down 40%, what is the present price?

12. A man whose wages are $4 a day spends $12\frac{1}{2}$% of his wages for rent, 40% for food, 5% for fuel, and 20% for clothing. How much has he left from his week's wages?

13. A farmer raised 250 bushels of potatoes per acre on 8 acres. He sold 10% of the crop for 42 cents a bushel, 25% at 56 cents a bushel, and the remainder at 65 cents a bushel. How much did he get for the entire crop?

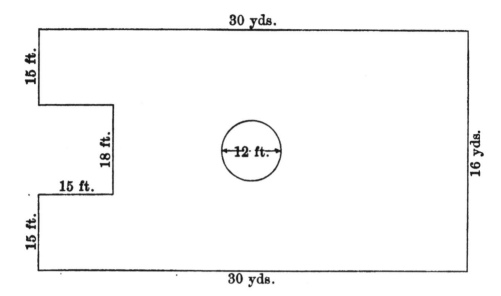

1. The figure represents a lawn containing a circular flower bed. What is the circumference of the flower bed?

2. How many plants can be set 15 in. apart around the border of the bed?

3. How many square feet are there in the surface of the flower bed?

4. How many plants could be set upon the whole bed if each plant should occupy not more than 2 sq. ft.?

5. After deducting the area of the flower bed, how many square yards of grass are there in the lawn?

6. How much would it cost to care for the lawn, aside from the flower bed, at the rate of 25 cents a square rod?

7. How many posts will it take for a picket fence across the north side, the east side, and the south side, if the posts are to be set not more than 12 ft. apart?

8. How many pickets will it take to go around the 3 sides, if the pickets are 2 in. wide, and the spaces between them are 2 in.?

Drill Work

Form as many examples as desired by adding parts of the columns.

Add the following:

1.	2.	3.	4.
$846.56	$162.74	$604.75	$524.31
531.14	83.28	38.37	276.19
785.50	70.95	217.81	85.07
34.08	919.93	79.42	830.21
79.37	258.32	850.50	764.88
90.24	85.64	33.94	70.82
258.06	90.75	103.48	323.17
76.27	739.84	736.11	861.53
4.35	43.96	86.64	602.76
583.52	152.60	98.35	53.35
16.53	75.34	355.09	42.83
433.47	641.50	72.43	798.60
24.98	284.65	295.93	362.27
80.64	16.78	70.17	30.75
937.39	658.77	125.82	456.87
59.78	785.34	67.58	73.13
769.18	940.49	793.26	144.06
141.35	46.83	408.31	608.32
54.36	432.29	95.49	46.25
376.43	51.36	14.63	764.36
467.35	200.54	236.88	18.87
59.43	374.95	56.51	216.34
347.66	956.28	94.04	108.65
891.38	88.73	221.38	793.15
40.23	797.86	73.50	46.54
746.35	40.23	891.43	80.61
54.70	496.68	660.75	727.45

Miscellaneous Problems

1. If 6 men can do a piece of work in 9 days, how long will it take 2 men to do it?

2. If 5 men can do a piece of work in 6 days, how many men can do it in 2 days?

3. If it costs $5 to build a piece of fence at 20 cents a rod, how much would it cost at 30 cents a rod?

4. If the dividend is 153,900 and the quotient 475, find the divisor.

5. If the subtrahend is 2136 and the remainder 6289, what is the minuend?

6. If the multiplier is 87 and the product 40,281, what is the multiplicand?

7. If the sum of 3 numbers is 1779, and 2 of the numbers are 653 and 342, what is the third number?

8. What part of a mile is 20 ft.?

9. What part of an oblong 7 ft. × 6 ft. is an oblong 3 ft. × 2 ft.?

10. How many times does a wheel whose circumference is $8\frac{1}{2}$ ft. revolve in going 10 rods?

11. How many times does a wheel whose diameter is 28 in. revolve in going $\frac{1}{8}$ of a mile?

12. If the fore wheel of a carriage is 9 ft. in circumference and the hind wheel 10 ft. in circumference, how many more times does the fore wheel revolve than the hind wheel in going a mile?

13. If $\frac{3}{4}$ of the value of a house is $6000, what is its value?

14. If $\frac{3}{4}$ of the value of a farm is $9000, what is $\frac{2}{3}$ of its value?

1. A rectangular mirror is 28 in. long and 18 in. wide measured on the outside of the frame. The frame is 5 in. wide. What are the length and width of the glass inside the frame?

2. How many square inches are there in the frame?

3. A garden is surrounded by a walk 3 ft. wide. The length and width measured outside the walk are 25 ft. and 18 ft. What is the area of the garden?

4. What is the area of the ceiling of a room which is 14 ft. long and $18\frac{1}{2}$ ft. wide?

5. Find the area of the four walls of a room which is 15 ft. long, 12 ft. wide, and 9 ft. high.

6. How many tiles, each 4 in. square, will be required for a piece of tiling 4 ft. long and 2 ft. 4 in. wide?

7. A township is 6 miles square. Into how many farms $\frac{1}{4}$ of a mile square can it be divided?

8. Into how many townships can a tract of land 24 miles long and 18 miles wide be divided?

9. The area of a rectangle is 36 sq. ft. What might its length and width be?

10. If a man receives $15 a week for his work, and spends $$2\frac{1}{4}$ for each $5 that he receives, how much will he save in 12 weeks?

11. How should 30 pieces of candy be divided among 3 children so that as often as the first child receives 1 piece, the second will receive 2 pieces, and the third 3 pieces?

12. Divide $100 in the proportion of 2 and 3.

13. Divide $200 in the proportion of 1, 4, and 5.

Original Problems

Make problems and solve them :

1. The product of two numbers is 27.

2. The product of three equal numbers is 125.

3. The product of three numbers is 60, and the product of two of them is 15.

4. The area of a rectangle is 72 feet, and it is 8 feet wide.

5. A cube contains 64 cubic inches.

6. I sold goods which cost me $4.00 at a gain of $12\frac{1}{2}\%$.

7. A fruit dealer bought oranges at $2.50 a box. Four boxes decayed, and he sold the remainder at $3.50 a box.

8. There are 231 cubic inches in a gallon.

9. Goods which cost $6 are sold for $7.

10. A produce dealer, having a car load of potatoes, sold $37\frac{1}{2}\%$ and 25% of them.

11. A bill of goods amounting to $450.50 is discounted 20%.

12. A dealer, having 240 barrels of flour, sold $12\frac{2}{3}\%$ of it to one man, 25% to another, and $16\frac{2}{3}\%$ to another.

13. A farmer who raised 1000 bushels of corn sold $\frac{2}{5}$ of it for 40 cents a bushel.

14. A picket fence is to be built around a garden which is 40 feet long and 25 feet wide.

15. A garden is surrounded by a walk 4 feet wide.

16. The area of a rectangle is 56 square feet.

17. A tank of water is 12 feet deep.

18. A dam across a stream is 30 feet long and 10 feet deep.

185

Review Problems

1. How many dozen are there in a gross?

2. How many dozen are there in 6 dozen dozen?

3. How much will 20 cases of eggs be worth at 16 cents a dozen if each case contains 49 doz.?

4. How many half gross boxes of pencils will it take to supply a school of 640 pupils for a year if each pupil is allowed 3 pencils a year?

5. How many sheets of paper are there in a ream?

6. How many quires of paper will a lady use in a year if she writes 1 letter every week day and 2 letters every Sunday, and uses 2 sheets of paper for each letter she writes?

7. How many reams of essay paper must a teacher order to last a school of 200 pupils for a term of 16 weeks, if the pupils average 4 sheets for each essay and write essays once in 2 weeks?

8. How much shall I gain by buying 200 barrels of flour at $4.25 a barrel and selling it at a gain of 15 per cent?

9. If I should pay $80 for goods and should sell them for $90, what part of the cost would the gain equal?

10. What would be the gain per cent?

11. In a rectangular garden 12 yd. long and 8 yd. wide there are 2 circular flower beds, each 14 ft. in diameter, and 4 oblong beds, each $12\frac{1}{2}$ ft. by 6 ft. Find the area of the circular beds.

12. Find the area of the oblong beds.

13. Find the area of that portion of the garden which is not occupied by the beds.

186

Percentage

1. 20 is $\frac{1}{10}$ of what number?
2. 40 is $\frac{1}{8}$ of what number?
3. 18 is $\frac{2}{3}$ of what number?
4. What is 10% of 20?
5. 25 is 10% of what number?
6. 3% of a number is 9. What is the number?
7. 6 is .03 of what number?
8. 48 is 12% of what number?
9. What is $12\frac{1}{2}$% of 48?
10. What is the number of which 80 is 50%?
11. What is the number of which 20 is 200%?
12. 24 is $12\frac{1}{2}$% of what number?
13. Find $12\frac{1}{2}$% of 56.
14. 4 is $33\frac{1}{3}$% of what number?
15. What is the number of which 12 is $16\frac{2}{3}$%?
16. What is $16\frac{2}{3}$% of 72?
17. If 6 is $12\frac{1}{2}$% of a number, what is $16\frac{2}{3}$% of the number?
18. If 24 is 6% of a number, what is $33\frac{1}{3}$% of the number?
19. If I have a number of oranges and sell 90% of them, what per cent have I left?
20. If I sell 90% and have 4 left, how many had I at first?
21. 6 pupils are $12\frac{1}{2}$% of the number in a room. How many are there in the room?
22. 5% of all the pupils in school would be 20. How many pupils are there?
23. What would be $33\frac{1}{3}$% of 1200 lb. of wool?

187

Compound Quantities

1. What part of a yard are 2 ft. 4 in. ?
2 ft. 4 in. = 28 in.   1 yd. = 36 in.
2 ft. 4 in. is $\frac{28}{36}$ or $\frac{7}{9}$ of a yard.

Find what part the one quantity is of the other:

2. 3 qt. 1 pt. of a gallon.

3. 3 pk. 5 qt. of a bushel.

4. 1 ft. 8 in. of a yard.

5. 4 ft. 2 in. of 7 ft. 6 in.

6. 5 sq. ft. of a square yard.

7. 144 cu. in. of a cubic foot.

8. 1 lb. 4 oz. of 2 pounds.

9. 2 lb. 4 oz. of 3 lb. 2 oz.

Multiply:

10. 19 bu. 3 pk. 5 qt. by 6.

11. 21 ft. 9 in. by 12.

12. 17 yd. 2 ft. 10 in. by 5.

13. 3 mi. 28 rd. 5 yd. by 10.

14. 18 sq. rd. 32 sq. ft. by 25.

15. 45 cu. ft. 1250 cu. in. by 25.

16. 64 lb. 12 oz. by 30.

How many times is the one contained in the other:

17. 5 pk. in 10 bu. 5 pk. ?

18. 2 qt. in 4 gal. 2 qt. ?

19. 3 qt. 1 pt. in 6 gal. 3 qt. 1 pt. ?

20. 6 in. in 12 ft. 6 in. ?

21. 10 in. in 2 yd. 2 ft. 8 in. ?

22. 4 rd. 12 ft. in a mile ?

23. 500 cu. in. in 4 cu. ft. 240 cu. in. ?

188

Specific Gravity

The weight of a substance compared with the weight of pure water is called its **specific gravity**. If a substance weighs twice as much as the same quantity of water, its specific gravity is 2. A substance which weighs 6.8 times as much as water has a specific gravity of 6.8.

The following table gives approximately the specific gravity of several substances:

Gold . . .	19.25		Cork . . .		.24
Lead . . .	11.35		Mercury . .		13.59
Silver . .	10.47		Milk . . .		1.03
Iron (cast) .	7.2		Sea water .		1.02
Marble . .	2.83		Pure water .		1.00

1. A cubic foot of water weighs 1000 ounces. Find the weight of a cubic foot of iron.

2. Find the weight of a cubic inch of gold.

3. How many cubic feet of marble will it take to weigh as much as a cubic foot of gold?

4. Find the weight of a cubic foot of milk.

5. How many more ounces does a cubic foot of milk weigh than a cubic foot of fresh water?

6. How many more ounces does a cubic foot of sea water weigh than a cubic foot of fresh water?

7. If a cubic foot of copper weighs 8780 ounces, what is its specific gravity?

8. A cubic foot of tin weighs 7290 ounces. What is the specific gravity of tin?

Percentage

1. Change to hundredths : $\frac{1}{2}$; $\frac{1}{4}$; $\frac{1}{8}$; $\frac{3}{8}$; $\frac{1}{6}$; $\frac{1}{10}$.

2. Change to hundredths : $\frac{1}{5}$; $\frac{3}{10}$; $\frac{7}{20}$; $\frac{2}{3}$; $\frac{5}{6}$; $\frac{3}{8}$.

3. 3 is what part of 4 ?   Change $\frac{3}{4}$ to hundredths.

4. 5 is what part of 6 ?   Change $\frac{5}{6}$ to hundredths.

5. 1 is how many hundredths of 4 ?   5 of 20 ?

6. 10 is what per cent of 20 ?   12 of 36 ?

7. 9 is what per cent of 45 ?   30 of 200 ?

8. What part of 10 is 5 ?   What per cent of 10 is 5 ?

9. When cloth which cost $4 a yard is sold for $3 a yard, what part of the cost is lost ?   What per cent is lost ?

10. If I buy a horse for $200 and sell him for $240, what part of the cost do I gain ?   What per cent do I gain ?

> I gain $40; $40 is $\frac{40}{200}$ or $\frac{20}{100}$ of the cost.   I gain 20%.

11. If I sell goods for $3.15 which cost $3, how many hundredths of the cost do I gain ?

12. If I buy goods for 50 cents, and sell them for 45 cents, what per cent of the cost do I lose ?

13. A merchant buys flour at $5 a barrel, and sells it at $4.50 a barrel.   What per cent does he lose ?

14. A merchant sells cloth for $37\frac{1}{2}$ cents a yard which cost 50 cents a yard.   What per cent does he lose ?

15. In a school of 60 pupils 12 pupils are absent. What per cent are absent ?

16. If upon the average 3 pupils are tardy each day, what per cent are tardy ?

17. In a town having a population of 4000 adult people there are 200 people who cannot read or write.   What is the per cent of illiterates ?

Percentage

1. 10 is $\frac{5}{100}$ of a certain number. What is the number?
2. 20 is 4% of what number?
3. 16 is 8% of what number?
4. 8 is 16% of what number?
5. What is 10% of 40?
6. 40 is 10% of what number?
7. What is 40% of 10?
8. 10 is 40% of what number?
9. What is the number of which 20 is $12\frac{1}{2}$%?
10. What is $12\frac{1}{2}$% of 20?

11. The Germans in a certain town number 1500, and they comprise 30% of the population. What is the population of the town?

12. In a town which has 12,600 inhabitants $16\frac{2}{3}$% are foreign born. How many are native born?

13. A clergyman saves 30% of his salary, and his savings amount to $1350 in 3 years. How large is his salary?

14. If I expend 80% of my salary, and have $240 left, how large is my salary?

15. About 75% of the weight of potatoes is water. How many pounds of dry matter are there in a bushel of potatoes weighing 60 lb.?

16. If $12\frac{1}{2}$% of milk is solids, how many ounces of solids would remain after evaporating 5 lb. of milk?

17. About $3\frac{1}{2}$% of average milk is butter fat. How many pounds of butter fat are there in 500 lb. of milk?

18. How many pounds of milk will it take to yield 14 lb. of butter fat?

19. If 4% of the milk of a certain cow is butter fat, how many pounds of milk must the cow yield during the year to produce 300 lb. of butter fat?

Fractions — Drill Work

See pages 25, 43, and 120.

1. $\frac{2}{3} + \frac{3}{4} + \frac{5}{6}$.
2. $\frac{1}{6} + \frac{5}{9} + \frac{1}{2}$.
3. $\frac{7}{8} + \frac{1}{8} + \frac{11}{12}$.
4. $\frac{1}{4} + \frac{5}{7} + \frac{1}{2}$.
5. $\frac{3}{5} + \frac{1}{6} + \frac{13}{15}$.
6. $\frac{3}{4} + \frac{1}{9} + \frac{11}{18}$.
7. $\frac{4}{5} + \frac{3}{8}$.
8. $\frac{6}{7} + \frac{1}{6}$.
9. $\frac{3}{8} + \frac{7}{16}$.
10. $\frac{1}{3} + \frac{18}{19}$.
11. $\frac{4}{5} + \frac{1}{40}$.
12. $\frac{1}{9} + \frac{1}{8}$.
13. $7\frac{3}{11} + 4\frac{3}{4}$.
14. $9\frac{4}{8} + \frac{5}{7}$.
15. $\frac{11}{12} + 10\frac{4}{5}$.
16. $\frac{2}{3} + 4\frac{11}{25}$.
17. $14\frac{1}{17} + 21\frac{1}{4}$.
18. $26\frac{1}{3} + \frac{15}{22}$.

19. $\frac{9}{10} - \frac{1}{2}$.
20. $\frac{11}{12} - \frac{5}{8}$.
21. $\frac{13}{15} - \frac{1}{2}$.
22. $\frac{23}{24} - \frac{3}{8}$.
23. $\frac{2}{5} - \frac{1}{7}$.
24. $\frac{3}{4} - \frac{2}{9}$.
25. $12 - 7\frac{5}{8}$.
26. $25\frac{11}{12} - 19$.
27. $28\frac{1}{2} - 8\frac{1}{4}$.
28. $32\frac{1}{4} - 10\frac{3}{4}$.
29. $35\frac{1}{3} - 12\frac{5}{6}$.
30. $40 - 17\frac{8}{11}$.
31. $\frac{1}{2} - \frac{7}{40}$.
32. $9\frac{7}{8} - \frac{11}{12}$.
33. $16\frac{5}{9} - 4\frac{7}{8}$.
34. $36 - \frac{13}{32}$.
35. $29 - 5\frac{3}{41}$.
36. $35\frac{1}{9} - 16\frac{4}{5}$.

37. $\frac{2}{8} \times 3$.
38. $\frac{7}{8} \times 5$.
39. $\frac{3}{4} \times 4$.
40. $\frac{4}{5} \times 20$.
41. 28×47.
42. $32 \times \frac{5}{8}$.
43. $\frac{2}{3} \times \frac{3}{4}$.
44. $\frac{12}{13} \times \frac{5}{6}$.
45. $\frac{4}{5} \times \frac{10}{11}$.
46. $\frac{7}{8} \times \frac{24}{25}$.
47. $\frac{32}{33} \times \frac{7}{16}$.
48. $\frac{3}{20} \times \frac{40}{49}$.
49. $3\frac{1}{4} \times 8$.
50. $9 \times 5\frac{2}{5}$.
51. $\frac{1}{4} \times 8\frac{12}{17}$.
52. $21\frac{14}{17} \times \frac{6}{7}$.
53. $12\frac{4}{5} \times 15\frac{3}{4}$.
54. $37\frac{8}{9} \times 16\frac{1}{4}$.

55. $\frac{12}{13} \div 4$.
56. $16 \div \frac{1}{3}$.
57. $25 \div 2\frac{1}{2}$.
58. $\frac{17}{18} \div \frac{5}{18}$.
59. $\frac{7}{8} \div \frac{1}{16}$.
60. $\frac{8}{25} \div \frac{1}{5}$.
61. $9\frac{3}{8} \div 2\frac{2}{3}$.
62. $14\frac{2}{7} \div 1\frac{1}{4}$.
63. $16\frac{1}{2} \div 8\frac{2}{5}$.
64. $\frac{8}{9} \div 5\frac{1}{4}$.
65. $17\frac{7}{11} \div \frac{1}{2}$.
66. $28 \div 4\frac{1}{7}$.
67. $\frac{36}{37} \div 9$.
68. $20 \div \frac{1}{5}$.
69. $24\frac{3}{5} \div 4\frac{1}{3}$.
70. $\frac{27}{38} \div \frac{5}{19}$.
71. $43\frac{2}{3} \div 10\frac{1}{4}$.
72. $62\frac{1}{7} \div 2\frac{2}{5}$.

192

Miscellaneous Problems

1. How much will it cost for wire enough to make 3 rows around a lot 10 rd. long and 4 rd. wide, if the wire costs 5½ cents a rod?

2. If a man steps 3 ft. at a step, and takes 100 steps in going around a square field, what is the length of the field?

3. If 3 qt. and 1 pt. of milk are taken from a can containing 3½ gal., how much will remain?

4. If a man can row a boat 4 miles an hour in still water, how far can he row in 6 hours down a river which flows 2½ miles an hour?

His rate down the river is 4 miles + 2½ miles an hour.

5. If a man can row 5 miles an hour in still water, and the river flows 2½ miles an hour, how long will it take him to row up the river a distance of 30 miles?

His rate up the river is 5 miles − 2½ miles an hour.

6. How long will it take him to row back down the river to the point of starting?

7. If I spend ½, ⅓, and 1/12 of my money, what fraction of it have I left? If I have $6 left, how much had I at first?

8. If the difference between ⅝ and ⅜ of my money is $12, how much money have I?

9. By selling a horse for ⅛ more than the cost I gained $12. What was the cost?

10. By selling a horse for ⅛ less than the cost I received $49. What was the loss?

11. In 3 farms there are 75 A. 43¼ sq. rd., 94 A. 115 sq. rd., and 178 A. 150½ sq. rd. respectively. How much land is there in the 3 farms combined?

Review Problems

1. How many days are there from April 27 to June 5?

2. If June should begin on Tuesday, what day of the week would the first day of the following October be?

3. If August should begin on Saturday, on what day of the week would the following September begin?

4. How many weeks are there in a year, and how many days over?

5. If a year which is not leap year should begin on Sunday, on what day of the week would the next year begin?

6. Find the exact number of days from Washington's Birthday to Aug. 10.

7. Find the time from July 4, 1776, to May 10, 1828.

8. Find the time from Nov. 19, 1882, to June 12, 1892.

9. How many square yards are there in the floor of a room $16\frac{1}{2}' \times 14'$?

10. How many yards of carpet will it take to exactly cover a floor $18' \times 13\frac{1}{2}'$, if the carpet is a yard wide?

11. If the carpet is $\frac{3}{4}$ of a yard wide, what part of a square yard is there in a piece 1 yd. long?

12. How many times is $\frac{3}{4}$ of a square yard contained in 6 sq. yd.?

13. If the carpet is $\frac{3}{4}$ of a yard wide, how many yards of length will exactly cover a floor $15' \times 12'$?

14. If a garden $30' \times 20'$ has a walk around it 4 ft. wide, how far is it around the outside of the walk?

15. Find how many square feet there are in the walk.

16. How many square inches of glass are there in a mirror if it is $28'' \times 22''$ measured outside the frame, and the frame is $3''$ wide?

194

Profit and Loss

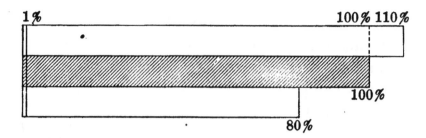

1. What number is 10% more than 100?

2. What number is 20% less than 100?

3. What number is 110% of 100?

4. What number is 80% of 100?

5. If 220 is 10% more than some number, what per cent is it of that number?

6. If 220 is 110% of some number, what is 1% of that number? What is 100% of the number?

7. If 240 is 20% less than some number, what per cent of that number is 240?

8. If 240 is 80% of some number, what is 1% of that number? What is 100% of the number?

9. John has 50% more marbles than William, and has 15 marbles. What per cent of William's number has he? How many marbles has William?

10. If I sell goods at a loss of 25% and receive $3, how much did the goods cost?

11. I sold a knife for 40 cents and lost 20%. What was the cost?

12. If I had gained 20%, for how much should I have sold it?

13. If I had sold the knife for 70 cents, what part of the cost should I have gained? What per cent should I have gained?

195

Miscellaneous Problems

1. There are 231 cubic inches in a gallon. How many gallons of water will it take to fill a box 14 in. long, 11 in. wide, and 6 in. deep?

2. How many cubic inches are there in a 10-in. cube?

3. What is the length of a cube which contains 125 cubic inches?

4. If the surface of a square is 64 sq. in., how long is it?

5. If a piece of board containing 8 sq. ft. is 6 in. wide, how long is it?

6. The surface of one side of a cube is 9 sq. in. How many cubic inches does the cube contain?

7. If a piece of board containing 12 sq. ft. is 12 in. wide at one end and 6 in. wide at the other, how long is it?

8. A can do a piece of work in 7 days. What part of it can he do in 2 days?

9. A can do a piece of work in 3 days, and B could do it in 6 days. What part of it can they both together do in 1 day?

10. How many days would it take A and B together to do the work?

11. A wagon load of earth is about a cubic yard. How many loads must be removed to make an excavation for a cellar 42 ft. long, 15 ft. wide, and $10\frac{1}{2}$ ft. deep?

12. The specific gravity of the air near the surface of the earth is about .0012. Find what would be the weight of the air in a room 22 ft. × 18 ft. × 8 ft.

13. The specific gravity of cork is .24. Find the weight of 10 cubic feet of cork.

196

Decimals — Drill Work

See pages 115, 144, and 161.

Find the products:

1. $20.53 \times 45.$
2. $156 \times 3.709.$
3. $2.31 \times 51.9.$
4. $34.8 \times .837.$
5. $.548 \times 9.25.$
6. $17.56 \times .123.$
7. $.125 \times 3.15.$
8. $.231 \times .642.$
9. $.087 \times 5.81.$
10. $.093 \times .372.$
11. $.026 \times .058.$
12. $.396 \times .009.$

13. $3.456 \times 1000.$
14. $2000 \times 8.256.$
15. $7.234 \times 100.$
16. $.9305 \times 1000.$
17. $.6758 \times 5000.$
18. $5000 \times .005.$
19. $8000 \times .0008.$
20. $3500 \times .0175.$
21. $73.21 \times 4200.$
22. $3.1416 \times 38.5.$
23. $7.25 \times 3.1416.$
24. $8.305 \times 5.123.$

Find the quotient to the nearest thousandth:

25. $34.225 \div 25.$
26. $75.831 \div 16.$
27. $832.14 \div 23.$
28. $1.5608 \div 19.$
    $3815.2 \div 31.$
    $13214 \div 24.$
29. $6.2518 \div 3.5.$
32. $35.416 \div 4.7.$
33. $218.32 \div .31.$
34. $5.2637 \div .42.$
35. $2175.6 \div 5.8.$
36. $10.123 \div .32.$

37. $43.1602 \div 100.$
38. $213.182 \div 1000.$
39. $5143.12 \div 2000.$
40. $813.145 \div .01.$
41. $31.4162 \div .001.$
42. $5161.82 \div .005.$
43. $3.10201 \div 25.$
44. $51832.6 \div 400.$
45. $32.4365 \div 5000.$
46. $1854.61 \div .003.$
47. $21.3165 \div 259.$
48. $2135.68 \div 1.231.$

Original Problems

Make problems and solve them:

1. A tank 50 feet high is filled with water.

2. A flume at the bottom of a dam 12 feet deep has an opening of 3 square feet.

3. A farmer sold 80% of his eggs.

4. 15% of the pupils in the school would be 45.

5. $37\frac{1}{2}$% of a lot of wool would be 1500 lb.

6. 6 men can do a piece of work in 8 days.

7. The subtrahend is 468, and the remainder 256.

8. The multiplier is 25, and the product 64,500.

9. The circumference of a wheel is $9\frac{1}{4}$ feet.

10. $\frac{3}{8}$ of the value of a farm is $1500.

11. Cloth which costs $5 a yard is sold for $4 a yard.

12. A merchant buys flour at $4.00 a barrel and sells it at $4.50 a barrel.

13. In a school of 80 pupils 4 pupils are absent.

14. A man saves 40% of his salary and saves $600 a year.

15. $3\frac{1}{2}$% of the milk from a certain cow is butter fat.

16. A man expends $\frac{1}{5}$ of his money for clothing, $\frac{1}{6}$ of it for groceries, and $\frac{1}{2}$ of it for rent.

17. A man takes 300 steps in walking around a square field.

18. A man takes 200 steps in going around a field which is twice as long as it is wide.

19. The summer vacation began on Friday, June 30th, and continued 10 weeks.

20. I sold a cow for $40 and lost 20%.

21. Henry has 25% more marbles than James, and has 20 marbles.

Trapezoids

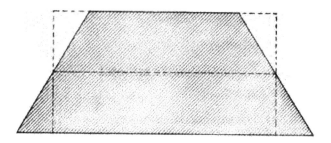

A **trapezoid** is a plane figure with four sides, having one pair of opposite sides parallel to each other. The parts of a trapezoid may be so arranged as to form a rectangle, whose length is midway between the lengths of the parallel sides.

Since the **area** of the trapezoid is the same as that of the rectangle, it equals the product of the altitude and the average length of the parallel sides.

1. What is the average between the numbers 18 and 24 ?

2. If the two parallel sides of a trapezoid are 12 ft. and 8 ft. long, what is the length of a rectangle equal to it ?

3. What is the area of a trapezoid whose altitude is 4 ft. 3 in., and whose parallel sides are 12 ft. and 18 ft. ?

4. Find the area of a trapezoid whose 2 parallel sides are 12 ft. 8 in. and 15 ft. 4 in., and the distance between whose parallel sides is 6 ft. 3 in.

5. From the top of a triangle whose base is 16 in. and altitude 12 in. a small triangle is taken off by a line parallel to the base, so that the altitude of the small triangle is 6 in. and its base 8 in. Find the area of the part remaining.

6. How many acres are there in a farm in the form of a trapezoid whose parallel sides are 120 rods and 164 rods, and whose width is 82 rods ?

7. The average between the 2 parallel sides of a trapezoid is 42 ft. 4 in., and the perpendicular distance between these sides is 9 ft. 6 in. What is the area of the trapezoid ?

Mental Problems

1. 12 is $\frac{1}{4}$ of what number?

2. 15 is $\frac{3}{4}$ of what number?

3. 15 is $\frac{1}{4}$ less than what number?
Why is this the same as Ex. 2?

4. 25 is $\frac{5}{4}$ of what number?

5. 25 is $\frac{1}{4}$ more than what number?

6. If $\frac{3}{8}$ of a number is 24, what is the number?

7. If $\frac{1}{8}$ less than a number is 30, what is the number?

8. If 25 is $\frac{5}{4}$ of a number, what is the number?

9. If 16 is $\frac{4}{5}$ of a number, what is the number?

10. If 9 is taken from 45, what part is taken away?

11. If 7 is added to 42, what part of 42 is added?

12. If 4 is taken from 10, what part of 10 is taken?

13. If 9 is added to 15, what part of 15 is added?

14. If $\frac{11}{10}$ of a number is 88, what is the number?

15. If $\frac{7}{8}$ of a number is 49, what is the number?

16. If I should increase my money by $\frac{1}{6}$ of it, I should have 28 cents. How much have I?

17. If I should spend 12 cents, I should have $\frac{4}{7}$ of my money left. How much have I?

18. A dealer sells lemons for $\frac{2}{4}$ more than they cost, and makes 6 cents a dozen. What is the cost?

19. If he sells lemons for $\frac{7}{9}$ of the cost, and loses 4 cents a dozen, what is the selling price?

20. $\frac{1}{12}$ of what I received for a book was gain. I gained 10 cents. How much was the cost?

21. I lost 15 cents by selling a book for $\frac{5}{8}$ of the cost. What was the cost?

Fractions

1. How many fourths are there in $5\frac{1}{4}$? $7\frac{3}{4}$?

2. How many wholes, with a remaining part, are there in $\frac{17}{3}$? $\frac{45}{7}$? $\frac{53}{25}$?

3. Change to improper fractions: $5\frac{4}{5}$; $7\frac{3}{8}$; $9\frac{5}{6}$; $25\frac{3}{7}$; $18\frac{5}{8}$.

4. Change to whole numbers, or mixed numbers: $\frac{25}{4}$; $\frac{63}{8}$; $\frac{96}{12}$; $\frac{531}{15}$; $\frac{257}{25}$.

5. If a man earns $\$12\frac{3}{4}$ in $5\frac{3}{8}$ days, how much does he earn per day?

6. A man exchanged $15\frac{3}{4}$ bu. of apples at $\frac{3}{8}$ of a dollar a bushel for meat at 15 cents a pound. How much meat did he receive?

7. If $13\frac{3}{4}$ tons of coal cost $\$89\frac{3}{8}$, what is the price per ton?

8. How many pounds of beef at $7\frac{3}{8}$ cents a pound can be bought for $\$100$?

9. If $29\frac{3}{4}$ bushels of wheat cost $\$21\frac{1}{4}$, how much will one bushel cost?

10. A cistern holds $224\frac{3}{4}$ gallons. How long will it take to fill it with water running through a pipe at the rate of $3\frac{5}{6}$ gallons a minute?

11. If 9 pounds of sugar cost $47\frac{1}{4}$ ¢, how much is that per pound?

12. How much would $23\frac{1}{2}$ pounds cost at the same rate?

13. A man walked $21\frac{3}{8}$ miles in $4\frac{3}{4}$ hours. What was the average number of miles per hour?

14. At the rate of $3\frac{5}{6}$ miles per hour, how far would a man walk in $7\frac{1}{6}$ hours?

201

Review

Express the following fractions as per cent:

1. $\frac{1}{4}$; $\frac{1}{5}$; $\frac{3}{5}$; $\frac{3}{10}$; $\frac{4}{5}$; $\frac{7}{10}$; $\frac{3}{20}$; $\frac{7}{25}$; $\frac{24}{25}$.

2. $\frac{1}{8}$; $\frac{2}{3}$; $\frac{1}{6}$; $\frac{5}{6}$; $\frac{1}{8}$; $\frac{7}{8}$; $\frac{1}{12}$; $\frac{5}{12}$.

What common fractions are equivalent to the following per cent?

3. 20%; 40%; 70%; 90%; 25%; 45%.

4. $12\frac{1}{2}\%$; $37\frac{1}{2}\%$; $33\frac{1}{3}\%$; $66\frac{2}{3}\%$; $6\frac{1}{4}\%$; $87\frac{1}{2}\%$.

5. 25 is what per cent of 125?

6. 36 is what per cent of 18?

7. 9 is what per cent more than 8?

8. 10 is what per cent less than 12?

9. What is 8% of 300?

10. 6 is what per cent of 300?

11. If 3 men can build 10 rods of wall in 2 days, how much will they build in 5 days?

12. If a quantity of provisions will last 4 men 8 days, how long will it last one man? How long will it last 6 men?

13. If 5 men can do a piece of work in 10 days, how many men will it take to do it in 2 days?

14. Two men leave the same place and travel in opposite directions, one $3\frac{1}{2}$ miles an hour, and the other $3\frac{3}{4}$ miles an hour. How far will they be apart at the end of 3 hours?

15. Two men at places 50 miles apart start at 7.30 A.M. to travel toward each other, one traveling $3\frac{1}{4}$ miles an hour, and the other $4\frac{1}{4}$ miles an hour. How far will they be apart at noon?

Construction

Review pages 92, 123, 151, and 199.

1. Construct a parallelogram having an angle of 100°. Find its altitude and area.

2. Construct an isosceles triangle. Find its altitude and area.

3. Construct an equilateral triangle. Find its altitude and area.

4. Inscribe a circle in a square 3½ inches long. Find the circumference of the circle. Find the area of the circle.

5. Draw a horizontal line 4 inches long. Above the line, at one end, make an angle of 70°, and at the other end an angle of 60°. Prolong the sides of the angles. On the side of the angle of 60° measure 2 inches from the angle. At this point, and on the same side as the angle, make an angle of 120°. Prolong the side, completing the figure. Which two angles together make 180°? Measure the sides and the altitude of the trapezoid and find its area.

6. Construct a trapezoid with one of the parallel sides 4½ inches long. Find the altitude and area.

7. Construct an isosceles triangle with a base line 4 inches long and the angles at the base each measuring 40°. Draw a line from the vertex to the base dividing the angle at the vertex into two equal angles. Measure the angles formed by this line at the base. Measure the parts into which the base is divided.

8. Construct an equilateral triangle with a side 4 inches long. Draw a line connecting the vertex with the middle point of the base. Measure the two angles formed at the vertex. Measure the two angles formed at the base.

203

Miscellaneous Problems

1. How far will a boy ride a bicycle in an hour at the rate of a mile in 4 minutes?

2. How far will a train go in an hour at the rate of 2 miles in 3 minutes?

3. How long will it take a man to walk 250 miles, if he walks $3\frac{1}{2}$ miles an hour, 8 hours a day, and rests on Sunday?

4. Find the length of 1° of a circle whose circumference is 8000 inches.

5. What would be the length of 12° on a great circle of a sphere whose circumference is 240 ft.?

6. If 4% of the milk of a cow is butter fat, how many pounds of butter fat will there be in the week's milk of a cow which yields 22 lb. of milk a day?

7. If $\frac{9}{10}$ of the weight of ordinary butter is pure fat, how many pounds of butter may be made from the week's milk?

8. Express as decimals: 25%; 5%; 110%; 255%.

9. Express as per cent: .15; 1.5; 15; 3.07.

10. The population of a city in 1892 was 42,000. It increased 5% each year for 2 years following. What was the population in 1894?

11. If the capital of a man in business is $12,000, and he increases it 10% each year for 3 years, how much capital will he then have invested?

12. Pure water is composed of 88.9% oxygen and 11.1% hydrogen. How many pounds of each are there in a cubic foot of water?

13. Find how many pounds of oxygen there are in a tank of water 6 ft. × 3 ft. × 2 ft.

204

Percentage

1. If I pay $5.00 a barrel for flour and sell it at a gain of 10%, how much shall I gain per barrel?

2. If I pay $5.00 and sell for $6.00, what per cent shall I gain?

3. What per cent of 200 is 40?

What part of 200 is 40? Reduce this fraction to per cent.

4. What per cent of 150 is 18?

5. If I pay $5.00 for flour and sell at a loss of 20%, how much shall I receive?

6. If I pay $5.00 and sell for $3.00, what per cent do I lose?

I lose $2; $2 is what per cent of $5?

7. If I make 12 cents a pound by selling tea at a gain of 20%, how much did the tea cost?

12 cents gain is a 20% gain; 12 cents is 20% of what number?

8. If I sell raisins for ⅔ of what they cost, what per cent do I make?

9. 12 is ⅕ more than what number?

This is the same as the problem, 12 is ⅚ of what number?

10. 12 is 120% of what number?

11. 30 is what per cent more than 25?

How many more than 25 is 30? 5 is what per cent of 25?

12. 30 is what per cent less than 40?

13. If I sell raisins at 12 cents a pound and gain 20%, what was the cost?

This is the same as Ex. 9. Why?

14. If I lose $\frac{1}{10}$ of the cost of an article, for what per cent of the cost do I sell it?

205

Review Problems

1. If a man travels away from home 9 hours, at the rate of 6 miles an hour, how long will it take him to return at the rate of $4\frac{1}{2}$ miles an hour?

2. At a place where the river flows $3\frac{3}{4}$ miles an hour, how far can a man row down river in $3\frac{5}{8}$ hours, if he can row $5\frac{1}{4}$ miles an hour in still water?

3. How far can he row up river in $4\frac{1}{2}$ hours?

4. How long would it take him to row 9 miles down the river and back to the point where he started?

5. How much shall I make on 3 gross of pens, if I buy them at 65 cents a gross and sell them at the rate of 6 for 5 cents?

6. If a crate of strawberries contains 2 doz. boxes, how much shall I make in buying 10 crates at $1.50 a crate and selling them at 3 boxes for a quarter?

Find what per cent of change is made in changing the prices of articles as follows:

7. Hats from $2.00 to $2.20.

8. Gloves from $1.25 to $1.50.

9. Shoes from $4.00 to $3.60.

10. Trousers from $2.50 to $3.00.

11. Collars from 12¢ to 10¢.

12. Neckties from 50¢ to 35¢.

13. If I should lose $\frac{1}{4}$ of the cost by selling goods for $3.00, what was the cost?

14. For how much should I have sold these goods to gain $\frac{1}{4}$ as much as the cost?

15. If I lose 25% by selling goods for 60¢, what was the cost?

206

Decimal Fractions

Read the following:

1. .3; .27; .256; .008; .1725.

2. 5.7 lb.; 9.42 in.; .759 T.; .04 sq. mi.; 14.25 A.

Write in figures:

3. Twenty-five and eight tenths; four thousandths; fifty-seven thousandths; twenty-two and sixteen ten thousandths.

4. Seven tenths of a rod; thirty-two thousandths of a mile; five and six tenths pounds; six hundred forty-two thousandths of a ton.

5. Add two hundred fifty-three and nine hundredths; sixty-four and eight ten thousandths; twenty-nine hundredths; two thousand five hundred three and six tenths; ninety-six and seventeen thousandths.

6. From nine and fifteen thousandths take three and fifteen ten thousandths.

7. How much will 22.675 T. coal cost at $8.25 a ton?

8. How many times is 5.21 miles contained in 243.9843 miles?

Add:

9.	10.	11.	12.
.85	2.8	52.4	25
8.13	.5	.0786	3.01
.084	5	8.3	.165
.25	8.5	51	25.4
1.5	1.04	8.75	.23
435.5	34	.875	173.3
45.935	.074	92.5	5.64
.05	5.8	.24	.0564
725	509	8.56	17.25

Ratio

See pages 111 and 164.

1. What is the ratio of 8 to 2 ? 4 to $\frac{1}{2}$? $\frac{3}{4}$ to $\frac{1}{4}$? $3\frac{3}{8}$ to $\frac{1}{8}$?

2. What is the ratio of $5 to 50 cents ? $3.75 to 25 cents ? $100 to 10 cents ?

3. What is the ratio of 2 gal. to 1 pt. ? Of 3 bu. to 6 qt. ?

4. What is the ratio of 1 yd. to 5 in. ? A rod to 6 in. ?

5. What is the ratio of a 4-in. square to a 2-in. square ?

6. What is the ratio of a 4-in. cube to a 2-in. cube ? What is the ratio of 3 yd. 1 ft. to 1 ft. 6 in. ?

7. How many rectangles 4 in. × 3 in. can be formed by the division of a rectangle 15 in. × 12 in. ?

8. How many rectangular solids 4 in. × 3 in. × 2 in. can be formed by the division of a rectangular solid 12 in. × 9 in. × 8 in. ?

9. How many farms 160 rods long and 40 rods wide are there in a township 6 miles square ?

10. How many piles of wood 10 ft. × 6 ft. × 4 ft. are contained in a pile 20 ft. × 12 ft. × 9 ft. ?

11. Find the ratio of the floor surface to the window surface of a room 32 ft. × 28 ft., which has 3 windows each containing 12 panes of 30 in. × 22 in. glass.

12. The area of New York State is 49,170 sq. miles. Its population in 1900 was 7,268,009. What was the ratio of its population to its area in square miles ?

13. The length of the St. Lawrence River is 2000 miles, and the area of the basin which it drains is 350,000 sq. miles. What is the ratio of its basin in square miles to its length in miles ?

208

The Air

If we disregard certain other substances which exist in the atmosphere in minute quantities, we may consider the following as its average composition. The figures indicate proportional quantities and not weights.

Nitrogen . . . . . . 79.02 %.
Oxygen . . . . . . 20.94 %.
Carbonic Acid . . . . .04 %.
100.00 %.

1. Find how many cubic inches of oxygen there are in a cubic foot of air.

2. How many cubic feet of air are there in a room which is $15' \times 12' \times 9\frac{1}{2}'$?

3. How many cubic feet of oxygen would there be in this room if the air were fresh ?

4. If a child inhales 25 cubic inches of air at each breath, how many times would he breathe to inhale a cubic foot of air ?

5. If a child breathes 20 times a minute, about how long does it take him to inhale a cubic foot of air ?

6. If there are 40 children in a schoolroom, and they breathe 20 times a minute upon the average, and inhale 25 cubic inches of air at each breath, about how large a quantity of air do they all together inhale in a minute ?

7. If a schoolroom is 32 ft. × 30 ft. × 10 ft., how many cubic feet of air does it contain ?

8. If all the air which is exhaled could be made to pass off through the ventilator so as not to be inhaled a second time, how long would the room full of air last the children ?

9. How many cubic feet of air must be admitted to the room each minute in order to furnish a constant supply of fresh air equal to the amount that the children inhale ?

Miscellaneous Problems

1. Change to common fractions in their smallest terms: 50%; 125%; 200%; 350%; 500%.

2. Change to common fractions: $\frac{1}{2}$%; $\frac{1}{4}$%; $\frac{1}{8}$%; $\frac{1}{6}$%; $\frac{2}{5}$%.

3. Change to per cent: $\frac{1}{6}$; $\frac{1}{4}$; $\frac{1}{8}$; $\frac{3}{8}$; $\frac{2}{8}$; $\frac{5}{6}$; $\frac{7}{8}$; $\frac{9}{10}$.

4. Find the number of strips of paper upon the walls of a room which is 20 ft. long and 16 ft. wide, if the paper is 18 in. wide.

5. If the room is 9 ft. high, what would be the entire length of all the strips of paper necessary to cover the walls, if no allowance is made for doors, windows, etc.?

6. What will be the cost of the paper at $.40 a roll, if each roll consists of a strip 24 feet long?

7. How much will it cost to paper a room 17 ft. long, 13 ft. wide, and 10$\frac{1}{2}$ ft. high with paper 18 in. wide and 24 ft. long at $.35 a roll, if the doors, windows, etc., make a difference of 15 yd. in length of paper?

8. How much will it cost to carpet a room 28 ft. long and 24 ft. wide with carpet $\frac{3}{4}$ of a yard wide at $1.00 a yard, if the strips run lengthwise and an allowance of 1 ft. extra for each strip is made for matching?

9. On what day of the week will a man finish a journey of 942 miles, if he starts on Monday morning and travels 12 miles an hour, 8 hours a day, and 6 days in a week?

10. On what day of the month will a boy reach his destination, if he has a thousand miles to ride upon a bicycle and starts on the morning of Monday, June 24th, riding 6 days in a week, 7 hours a day, and 9 miles an hour?

Mental Problems

1. What per cent of a gallon is 1 qt.? 3 qt.? 1 pt.? 3 pt.?

2. What per cent of a yard is 1 ft.? 2 ft.? 6 in.? 30 in.?

3. What per cent of a dime is 1 cent? 3 cents? 6 cents? 9 cents?

4. A merchant buys goods for $12 and sells them for $15. What per cent does he gain?

5. I have $4 in one pocket, which is 25% of what I have in the other. How much have I in all?

6. A lady had $36 and spent 33⅓% of it. How much had she left?

7. If I buy goods for $16 and sell them for $18, what per cent do I make?

8. After spending 30% of my money I had $14 left. How much had I at first?

9. What must be the selling price in order to gain 15% on goods which cost 40 cents?

10. If I buy for $15 and sell for $12, what per cent do I lose?

11. If I buy for $18 and sell so as to make 16⅔%, how much do I make?

12. If I sell goods for $20 which cost $24, what part of the cost do I lose?

13. Find how much I shall gain by making 12½% on goods which cost $248.

14. Find how much I shall lose by selling goods which cost $300 at a loss of 7%.

15. How much shall I receive for goods which cost $600, if I sell them at a gain of 8%?

211

Percentage

1. Find 6% of 2543.

1 per cent of 2543 is $\frac{1}{100}$ of 2543, which is 25.43. 6 per cent of 2543 is 6 times 25.43, which is 152.58.

2. Find $\frac{1}{2}$% of 248.

1 per cent of 248 is 2.48. $\frac{1}{2}$ per cent of 248 is $\frac{1}{2}$ of 2.48, or 1.24.

Find :

3. 3 per cent of 256.

4. 6 per cent of 783.

5. 9 per cent of 2530.

6. 13 per cent of 3562.

7. 17% of $1425.

8. 22% of $384.92.

9. 31% of $658.40.

10. 42% of $853.50.

11. Find the number of which $3.80 is 19%.

Since $3.80 is 19% of the number, 1% of the number is $\frac{1}{19}$ of $3.80, or $.20, and 100%, or the number itself, is 100 times $.20, or $20.

Find the number of which :

12. 25 is 5 per cent.

13. 64 is 8 per cent.

14. 81 is 9 per cent.

15. 154 is 7 per cent.

16. $1.47 is 7%.

17. $3.20 is 16%.

18. $16.80 is 8%.

19. $14.70 is 35%.

Find :

20. 1 per cent of 2642.

21. $\frac{1}{2}$ per cent of 2642.

22. $2\frac{1}{2}$ per cent of 2642.

23. $\frac{1}{4}$ per cent of 1364.

24. $5\frac{1}{8}$% of $53.60.

25. $7\frac{1}{2}$% of $156.42.

26. $8\frac{1}{4}$% of $387.20.

27. $12\frac{3}{4}$% of $895.46.

Find the number of which :

28. 200 is $12\frac{1}{2}$ per cent.

29. 300 is $16\frac{2}{3}$ per cent.

30. 561 is $33\frac{1}{3}$ per cent.

31. 50 is $\frac{1}{2}$ per cent.

32. 20 is $\frac{3}{4}$%.

33. 40 is $\frac{5}{8}$%.

34. 60 is $\frac{2}{3}$%.

35. 36 is $\frac{2}{5}$%.

Original Problems

Make problems in percentage based upon the following statements, and solve them:

1. Mary has 8 cents and Lizzie has 10 cents.
2. The cost of an article is $\frac{4}{5}$ of the selling price.
3. The cost of an article is $\frac{5}{4}$ of the selling price.
4. A farmer had 400 sheep, and lost 30% of them.
5. A farmer had 200 chickens and lost 25 of them.
6. 12 gal. of water is added to 36 gal. of alcohol.
7. A boy lost 40% of his marbles, and had 24 marbles left.
8. A watch which cost $40 is sold for $35.
9. $\frac{1}{8}$ of the pupils in a room are over 10 years old, and $\frac{1}{6}$ of them are over 11 years old.
10. A box contains 4 doz. oranges, but 8 oranges are spoiled.
11. I bought a horse for $150 and sold him for $175.
12. I bought a cow for $50 and sold her for $48.
13. A man saved $37\frac{1}{2}$% of his salary and saved $240.
14. Gold was once worth 30% more than currency.
15. I sold goods for $70, and lost 30%.
16. I sold goods at a gain of 25%, and received $45.
17. One field contains 10 sq. rd., and another is 10 rd. square.
18. In a schoolroom there are 20 girls and 16 boys.
19. Cloth is marked to sell at 60 cents a yard, and is afterwards sold at a discount of 15%.
20. Cloth is marked to sell at 40 cents a yard, and is afterwards sold at 30 cents a yard.
21. Goods are sold for $100 at a loss of $33\frac{1}{3}$%.
22. Goods are sold for $100 at a gain of 25%.

213

Cylinders

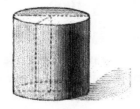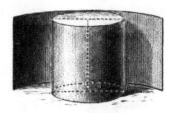

If the curved surface, or lateral surface, of a cylinder were spread out upon a plane it would become a rectangle. The circumference of the cylinder would become the length of the rectangle and the altitude of the cylinder the width of the rectangle.

1. Find the distance around a cylinder which is 7 in. in diameter and 20 in. long.

2. Find the surface of one end of such a cylinder.

3. What is the entire surface of the cylinder?

4. Find the distance around a pulley which is 14 in. in diameter.

5. Find the number of square inches on the outer surface of this pulley, if it is 7 in. wide.

6. If this pulley makes 6 revolutions while the belt is passing once around it, how long is the belt?

7. If the same belt goes around another pulley, which is 7 in. in diameter, how many more times will the small pulley revolve than the large pulley while the belt is going around once?

8. The cylinder of a street roller is 6 ft. in diameter and 8 ft. 6 in. long. Find the number of square feet upon its outer surface.

9. How many square feet of the street does it roll down in making one revolution?

10. How many square yards does it pass over in making 5 revolutions?

Easy Methods

1. Multiply 25 by 10.   By 100.   By 1000.

2. Divide 4320 by 10.   By 100.   By 1000.

In multiplying by any multiple of 10 we simply annex the number of ciphers contained in the multiplier, and in dividing by any multiple of 10 we place the decimal point before as many figures of the dividend as there are ciphers in the divisor.

3. Multiply 25.375 by 10.   By 1000.

4. Divide 6238.7 by 10.   By 1000.

In multiplying or dividing a number containing a decimal fraction by a multiple of 10 we move the decimal point to the right or left as many places as there are ciphers in the multiplier or in the divisor.

5. Multiply 9 by 100.   By 500.

500 equals 5 × 100.   100 times 9 equals 900, and 500 times 9 equals 5 times that, or 4500.

6. Divide 4800 by 100.   By 600.

600 equals 6 × 100.   100 is contained in 4800 48 times and 600 one sixth as many times, or 8 times.

Give the results orally :

7. $100 \times 275 =$	**17.** $30.56 \times 10 =$
8. $1000 \times 500 =$	**18.** $100 \times 3.1416 =$
9. $300 \times 23 =$	**19.** $457.3 \div 100 =$
10. $500 \times 250 =$	**20.** $8706 \div 1000 =$
11. $4000 \div 100 =$	**21.** $20 \times 25.04 =$
12. $18000 \div 300 =$	**22.** $70 \times 5.112 =$
13. $24800 \div 400 =$	**23.** $2460 \div 200 =$
14. $800 \times 800 =$	**24.** $315.6 \times 300 =$
15. $5000 \times 5000 =$	**25.** $.0034 \times 200 =$
16. $7200 \div 720 =$	**26.** $.0525 \div 500 =$

Review Problems

1. How long is a square which contains 25 sq. in. ?

2. How many miles is it around a square which contains 49 sq. miles ?

3. What is the square root of 100 ? 121 ? 169 ?
Find the square root mentally, by inspection.

4. What is the length of one side of a square which contains 225 sq. ft. ? 289 sq. ft. ?

5. If I buy melons for $1.44, paying as many cents per melon as there are melons, how many melons do I buy ?

6. What is the square root of 400 ? 1600 ? 3600 ?

7. How wide is a rectangle whose length is twice its width and which contains 200 sq. ft. ?
Think of the rectangle as composed of two squares.

8. How long is a rectangle which contains 100 sq. ft., and whose length is 4 times its width?

9. If a man breathes 18 times a minute and inhales 35 cubic inches of air at each breath, how many cubic feet of air does he inhale in an hour?

10. If we reckon 20% of the air as oxygen, how many cubic feet of oxygen does this amount of air contain?

11. If 25% of the oxygen of the air which is inhaled is retained in the system, how much oxygen is taken from the air every hour?

12. If about 8% of the breath exhaled is poisonous carbonic acid, how much carbonic acid is exhaled in an hour?

13. The fresh air comes into a certain schoolroom through a shaft whose area is 3 sq. ft. There are 45 children in the room. At the rate of how many feet a minute must the air move through the shaft in order that each child may have 2 cubic feet of fresh air per minute?

Multiplication of Fractions

1. What is one fourth of one half? Three fourths of one half?

2. What is one third of one half? Two thirds of one half?

3. What is one fourth of one third? One fourth of two thirds? Three fourths of one third? Three fourths of two thirds?

4. What is one fifth of one fourth? One fifth of three fourths? Three fifths of three fourths?

5. Multiply $\frac{3}{4}$ by $\frac{2}{3}$.

$\frac{3}{4}$ multiplied by $\frac{2}{3}$ is the same as $\frac{2}{3}$ of $\frac{3}{4}$. $\frac{1}{3}$ of $\frac{3}{4}$ is $\frac{1}{4}$. $\frac{2}{3}$ of $\frac{3}{4}$ is 2 times $\frac{1}{4}$, or $\frac{2}{4} = \frac{1}{2}$.

6. Multiply $\frac{4}{5}$ by $\frac{2}{3}$.

$\frac{1}{3}$ of $\frac{4}{5}$ is $\frac{4}{15}$. $\frac{2}{3}$ of $\frac{4}{5}$ is 2 times $\frac{4}{15}$, or $\frac{8}{15}$.

We obtain the same result by multiplying together the numerators and the denominators of the fractions. $\frac{2}{3}$ of $\frac{4}{5} = \frac{8}{15}$.

What is:

7. $\frac{1}{3}$ of $\frac{1}{6}$

8. $\frac{1}{3}$ of $\frac{5}{6}$

9. $\frac{2}{3}$ of $\frac{5}{6}$

10. $\frac{1}{4}$ of $\frac{3}{8}$

11. $\frac{3}{4}$ of $\frac{7}{8}$

12. $\frac{5}{6}$ of $\frac{4}{5}$

13. $\frac{2}{7}$ of $\frac{1}{2}$

14. $\frac{5}{6}$ of $\frac{3}{2}$

15. $\frac{8}{9}$ of $\frac{9}{10}$

Multiply:

16. $\frac{3}{8}$ by $\frac{3}{4}$

17. $\frac{3}{4}$ by $\frac{4}{5}$

18. $\frac{5}{8}$ by $\frac{1}{5}$

19. $\frac{6}{7}$ by $\frac{5}{13}$

20. $\frac{8}{9}$ by $\frac{14}{15}$

21. $\frac{9}{10}$ by $\frac{7}{11}$

22. $1\frac{1}{5}$ by $2\frac{1}{4}$

23. $2\frac{1}{8}$ by $3\frac{1}{5}$

24. $3\frac{3}{8}$ by $1\frac{1}{7}$

25. If a boy has $\frac{3}{4}$ of an apple and gives his sister $\frac{2}{5}$ of what he has, what part of the whole apple does he give her?

26. A dealer buys a quantity of fruit, and sells $\frac{1}{3}$ of it. $\frac{1}{4}$ of the remainder spoils. What part of the whole lot remains?

27. A man spends $\frac{2}{3}$ of his salary for board, and $\frac{1}{4}$ of what remains for clothing. How much has he left?

217

Right Triangles

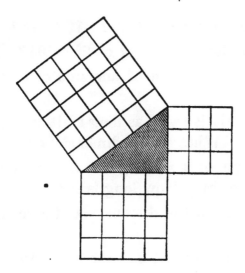

In a right triangle one of the two sides forming the right angle is called the **base** and the other the **altitude** or **perpendicular.** The side opposite the right angle is called the **hypotenuse.**

In the figure observe that the number of squares on the hypotenuse equals the sum of the squares of the other two sides.

In any right triangle the square of the hypotenuse is equal to the sum of the square of the sides. The square upon either side is equal to the difference between the square upon the other side and that upon the hypotenuse.

1. If the base of a right triangle is 8 in. long, how many square inches will there be in the square constructed on it?

2. If the altitude is 6 in., how many square inches will be contained in the square constructed upon it?

3. How many square inches will there be in the square upon the hypotenuse?

4. What will be the length of the hypotenuse?

5. What is the altitude of a right triangle whose hypotenuse is 13 ft. and base 12 ft.?

218

1. What is the square of 10 ? 12 ? 15 ? 25 ?

2. What is the square root of 25 ? 64 ? 121 ? 225 ?

3. What is the length of one side of a square which contains 256 sq. in.?

4. A rectangle is twice as long as it is wide, and contains 50 sq. in. What are its dimensions ?

5. The length of a rectangle is 5 times its width, and it contains 80 sq. in. How long is it ?

6. What may be the dimensions of a rectangle which contains 36 sq. in.?

7. What is the cube of 2 ? 3 ? 7 ? 10 ?

8. What is the cube root of 27 ? 64 ? 216 ?

9. What is the length of a cube which contains 125 cubic inches ?

10. What is the area of one end of a cube which contains 27 cubic inches ?

11. What is the entire surface of a cube which contains 216 cubic inches ?

12. The surface of one side of a cube is 16 sq. in. How many cubic inches does the cube contain ?

13. The entire surface of a cube is 150 sq. in. What is the length of its edge ?

Find the selling price of the following articles:

		Marked Price.	Discount.
14.	Woolens . . .	$1.25 . .	20%.
15.	Silks	1.50 . .	10%.
16.	Suits	28.00 . .	15%.
17.	Coats	12.00 . .	12½%.
18.	Hats	2.50 . .	16⅔%.
19.	Gloves	1.20 . . .	40%.

Percentage

1. Change $\frac{1}{12}$ to hundredths.

2. Change $\frac{1}{16}$ to hundredths.

3. What is $8\frac{1}{3}\%$ of 24 ?

4. What is $6\frac{1}{4}\%$ of 48 ?

5. What is $8\frac{1}{3}\%$ of 240 ?

6. What is $6\frac{1}{4}\%$ of 320 ?

7. 360 is $66\frac{2}{3}\%$ of what number ?

8. What number is $12\frac{1}{2}\%$ more than 64 ?

9. 36 is 50% more than what number ?

10. If I sell goods at a gain of $\frac{1}{12}$ of the cost, what per cent do I gain ?

11. If I sell goods so as to lose $\frac{1}{16}$ of the cost, what per cent do I lose ?

12. Find the cost of goods when the selling price is $15 and the gain $3. What is the gain per cent ?

13. Find the cost when the selling price is $20 and the loss is 20%.

14. $24 is $\frac{2}{5}$ of what a coat cost me. What was the cost ?

15. $60 is .15 of my indebtedness. How much do I owe ?

16. A boy sells a knife so as to make 40% of what it cost. He makes 10 cents. What was the cost ?

17. How much shall I gain by purchasing 20 bbl. of flour at $4.50 a barrel; 600 lb. sugar at $6\frac{1}{2}$ cents a pound; and 200 lb. of tea at 48 cents a pound, if I sell the whole at a gain of 15% ?

18. A merchant bought 240 yd. of cloth at $7\frac{1}{4}$ cents, and 60 shirts at 42 cents. He sold the cloth at a gain of 25%, and the shirts at a gain of $33\frac{1}{3}\%$. How much did he make in all ?

Cancellation

Review page 174.

1. Divide $2 \times 4 \times 3$ by $5 \times 8 \times 9$.

$$\frac{\cancel{2} \times \cancel{4} \times \cancel{3}}{\cancel{5} \times \cancel{8} \times \cancel{9}} = \frac{1}{15}$$

When the product of a number of factors is to be divided by the product of another number of factors, the division may be indicated in the form of a fraction. Equal factors may then be removed from the dividend and the divisor. This process is called **cancellation.**

2. What are the factors of 6? Of 12?

3. Express $\frac{6}{12}$ by writing the factors of 6 and 12 instead of the numbers themselves.

4. Change the fraction $\frac{6}{12}$ to its smallest terms by removing the equal factors.

5. Find the factors of 72. Of 36.

6. Divide 72 by 36 by removing equal factors from both.

7. Divide the product of 2, 3, 5, and 7 by the product of 2, 3, and 5.

8. Divide the product of 2, 2, 3, and 5 by the product of 2, 2, 3, 3, and 5.

9. Divide the product of 6, 8, 9, and 20 by the product of 8, 12, 15, and 30.

10. Multiply $\frac{2}{3}$ by $\frac{3}{4}$ by removing equal factors from the product of the numerators and the product of the denominators.

11. Multiply together $\frac{1}{2}$, $\frac{3}{4}$, $\frac{5}{6}$, $\frac{8}{9}$, and $\frac{3}{10}$.

12. Find $\frac{2}{3}$ of $\frac{5}{9}$ of $\frac{8}{15}$ of $\frac{9}{10}$.

13. How many times is a rectangular solid 7 ft. $\times$ 4 ft. $\times$ 3 ft. contained in a rectangular solid 21 ft. $\times$ 12 ft. $\times$ 6 ft.?

14. Multiply 1728 by 10, divide the result by 12, multiply the result by 5, and divide this result by 120.

Triangles

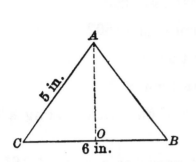 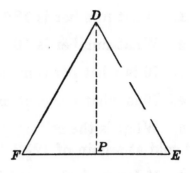

Review page 123.

In an isosceles triangle how does a line drawn from the vertex perpendicular to the base divide the triangle? What kind of triangles are formed by the division?

1. In an isosceles triangle the equal sides are each 5 in. long and the base is 6 in. long. Find the altitude of the triangle.

2. What is the area of this triangle?

3. If the base of an isosceles triangle is 10 ft. and the two equal sides are each 13 ft., what is the perpendicular distance from the base to the opposite vertex?

4. What is the area of such a triangle?

5. Find the altitude of an isosceles triangle whose base is 12 in. long and whose equal sides are each 10 in. long.

6. Find the area of the triangle.

7. In an equilateral triangle whose sides are each 6 ft. long the perpendicular is drawn from one vertex to the opposite side. How long is the base of one of the right triangles thus formed?

8. About what is the perpendicular distance from the vertex to the opposite side?

9. About what is the area of this triangle?

Miscellaneous Problems

1. What number is 18% less than 200 ?

2. What number is 16% more than 150?

3. 70 is what per cent less than 80 ?

4. 70 is what per cent more than 50 ?

5. What is the selling price of an article costing $32, and sold at a gain of $12\frac{1}{2}$% ?

6. If I gain $30 by selling goods at a gain of $16\frac{2}{3}$%, what was the cost of the goods ?

7. A merchant gained $40 by selling flour at a profit of 10%. It cost him $4 a barrel. How many barrels were there?

8. A merchant lost $25 by selling apples at a loss of $12\frac{1}{2}$%. They cost him $2 a barrel. How many barrels were there?

9. If $35 is divided between 2 men so as to give one $4 as often as the other is given $3, how much will each receive ?

10. A father divides his farm of 150 acres between his 2 sons in the proportion of 3 to 2. How many acres does each son receive?

11. If I buy an equal number of oranges and apples for 32 cents, and pay 5 cents for each orange and 3 cents for each apple, how many of each shall I buy ?

12. A butcher buys an equal number of cows, pigs, and sheep. Each cow costs $20, each pig $7, and each sheep $3. He pays $90 in all. How many of each does he buy ?

13. Three boys working together earn 96 cents. They divide the money in the proportion of 3, 5, and 8. How much does each boy receive ?

Common Fractions and Decimal Fractions

A decimal fraction may be changed to the form of a common fraction by writing the denominator and changing the fraction to lower terms when possible.

$$.25 = \tfrac{25}{100} = \tfrac{1}{4}$$

A common fraction may be changed to the form of a decimal fraction by changing it to tenths, hundredths, or thousandths, etc.

$$\tfrac{3}{20} = \tfrac{15}{100} = .15$$

Change to common fractions:

1. .5	6. .45	11. 8.5	16. 3.125
2. .25	7. .125	12. 9.35	17. 5.025
3. .15	8. .275	13. 12.14	18. 16.008
4. .24	9. .35	14. 15.06	19. 23.048
5. .17	10. .165	15. 22.08	20. 42.002

Change to decimal fractions:

21. $\tfrac{1}{2}$	26. $\tfrac{3}{20}$	31. $8\tfrac{1}{2}$	36. $.7\tfrac{1}{2}$
22. $\tfrac{1}{4}$	27. $\tfrac{4}{25}$	32. $9\tfrac{1}{4}$	37. $1.4\tfrac{3}{4}$
23. $\tfrac{3}{4}$	28. $\tfrac{7}{50}$	33. $12\tfrac{3}{4}$	38. $.04\tfrac{1}{4}$
24. $\tfrac{1}{5}$	29. $\tfrac{13}{20}$	34. $16\tfrac{1}{5}$	39. $8.31\tfrac{7}{10}$
25. $\tfrac{3}{5}$	30. $\tfrac{24}{25}$	35. $21\tfrac{3}{10}$	40. $9.1\tfrac{1}{2}$

Give the sums in the decimal form:

41.	42.	43.	44.
8.27	.137	$75\tfrac{4}{25}$	.005
.025	25.4	$11\tfrac{1}{4}$	$.026\tfrac{1}{2}$
$3\tfrac{3}{4}$	$9\tfrac{15}{20}$	142.05	$.15\tfrac{3}{5}$
$10\tfrac{1}{5}$	$15\tfrac{4}{5}$	81.017	$.013\tfrac{7}{10}$

Miscellaneous Drill Work

Divide by canceling the equal factors:

1. The product of 2, 5, 9 by the product of 4, 5, 6.
2. The product of 6, 12, 20 by the product of 12, 15, 10.
3. The product of 2, 3, 4, 5, 6 by that of 3, 4, 5, 6, 7.
4. The product of 7, 10, 16, 20 by that of 8, 14, 30.
5. The product of 11, 13, 34, 19 by that of 22, 17, 12.
6. The product of 25, 30, 55, 49 by that of 11, 35, 42.

Find measurements when a circle is inscribed in a square:

7. The diameter of a circle in a 12-inch square.
8. The area of a circle in a 12-inch square.
9. The area of the part of the square outside the circle.
10. The radius of a circle in an 18-inch square.
11. The area of a circle in an 18-inch square.
12. The area of the part of the square outside the circle.

Find the measurements of a cylinder six feet long:

13. The circumference, if the diameter is 4 feet.
14. The lateral surface, if the diameter is 4 feet.
15. The area of the ends, if the diameter is 4 feet.
16. The lateral surface, if the diameter is 16 inches.
17. The entire area, if the diameter is 16 inches.

Find the unknown side of the right triangle:

18. The base is 12 feet and the altitude 9 feet.
19. The hypotenuse is 13 feet and the base 12 feet.
20. The hypotenuse is 15 feet and the altitude 9 feet.
21. The altitude is 6 feet and the hypotenuse 10 feet.
22. The altitude is 12 feet and the base 16 feet.

Original Problems

Make problems and solve them:

1. The length of a rectangle is twice its width, and it contains 98 square feet.

2. A child breathes 22 times a minute, and inhales 28 cubic inches of air at each breath.

3. About 20% of the air is oxygen.

4. About 8% of the air exhaled from the lungs is carbonic acid.

5. A boy has ⅔ of an orange, and gives his sister ¾ of what he has.

6. The hypotenuse of a right triangle is 10 feet, and the altitude 6 feet.

7. The length of a rectangle is three times its width, and it contains 75 square inches.

8. The surface of one side of a cube is 36 square inches.

9. The entire surface of a cube is 54 square inches.

10. The marked price of suits was $16.00, but they are subject to a discount of 12½%.

11. The selling price of goods is $18.00 and the profit $3.00.

12. 48 is 50% more than some number.

13. In an isoseeles triangle a line is drawn from the vertex perpendicular to the base. One of the equal angles of the triangle is an angle of 50°.

14. The sides of an equilateral triangle are each 10 feet long.

15. The sum of the ages of a boy and his sister is 18 years.

16. A merchant gained $6 by selling goods for 20% more than the cost.

Prisms

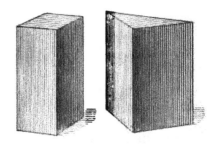

A **prism** has equal triangles, equal squares, or other equal figures for its two ends, and parallelograms for its side, or lateral faces. In a right prism the lateral faces are rectangles.

1. A right triangular prism has its 3 lateral faces equal. Each face is $18\frac{1}{2}$ in. long and 8 in. wide. What is the combined length of all the edges of the prism?

2. How many square inches are there in the 3 lateral faces?

3. If the altitude of each triangle at the ends is 6.9 in., how many square inches are there in the 2 ends?

4. What is the entire surface of the prism?

5. The perimeter of a triangle forming one end of a triangular prism is 16 ft. 8 in. The length of the prism is 9 ft. 6 in. What is the combined area of the 3 lateral faces?

6. What is the difference between the lateral surface of a regular triangular prism 12 ft. long, whose lateral faces are each 2 ft. 6 in. wide, and that of a square prism 10 ft. long and 2 ft. 6 in. square?

7. Find the entire surface of a square prism $4\frac{1}{2}$ ft. square and $8\frac{1}{2}$ ft. long.

8. What is the lateral surface of a cylinder 6 ft. long, whose diameter is 35 in.?

Percentage

1. What is $\frac{1}{100}$ of 600? 1% of 800? $\frac{1}{2}$% of 800? $\frac{1}{4}$% of 800?

2. What is $\frac{1}{100}$ of 340? 1% of 340? $\frac{1}{2}$% of 340? $\frac{1}{8}$% of 340?

3. $\frac{1}{100}$ of $24.80? $\frac{1}{2}$% of $24.80? $\frac{1}{4}$% of $32.40? $\frac{1}{5}$% of $65.50?

4. Express as decimals: $\frac{1}{4}$%; $2\frac{1}{2}$%; $25\frac{5}{8}$%.

5. Express as common fractions: 1%; $\frac{1}{2}$%; $\frac{1}{4}$%; $\frac{3}{4}$%.

6. Express as common fractions: $2\frac{1}{2}$%; $3\frac{1}{4}$%; $5\frac{1}{8}$%.

7. A shipment of 10,000 lb. of wool shrank $\frac{1}{4}$% in transportation. How much did it shrink?

8. A grocer gained $10 by selling a quantity of sugar at a gain of $\frac{1}{2}$%. What was the cost of the sugar?

9. If it costs $\frac{1}{5}$% to send money by express, how much will it cost to send $900?

10. How much is deducted from a bill of $250 by a discount of $\frac{1}{2}$%?

11. Find how much is gained by selling goods which cost $426.24 at a gain of $5\frac{3}{8}$%.

12. How much is saved by securing a discount of $\frac{1}{4}$% upon a bill amounting to $763.20?

13. If it costs $\frac{2}{5}$% of the value of a house to insure it for 1 year, how much will it cost to insure a house worth $3000?

14. A merchant bought 300 bbl. of pork, each containing 200 lb., at 7 cents a pound. He sold it at a gain of $4\frac{5}{8}$%. What was the amount of his profit?

15. If I make 50 cents by selling goods at a profit of 2%, what was the cost of the goods?

16. By selling cotton at a profit of $\frac{1}{4}$% a merchant gained $100. What was the cost of the cotton?

228

Commission

When one person transacts business for another he is called the **agent** of the other person. The agent usually charges for his services a certain per cent of the money expended or the money received. This is called his **commission**.

1. If an agent charges 5% commission, how much will he charge for selling goods to the amount of $100?

2. What would be the commission for selling 500 lb. of wool at 42 cents a pound, if the rate of commission is $\frac{1}{2}$%?

3. If the rate of commission is $2\frac{1}{2}$%, what will be the commission for selling goods to the amount of $342.50?

4. A broker bought for me a piece of real estate for $2350, and charged 1% commission. What was the amount of his commission? What was the entire cost of the real estate?

5. $2 is what per cent of $50?

How much is one per cent of $50?

6. If an agent's commission for selling goods at a commission of 2% is $4, how much did he get for the goods?

7. If an agent charges me $10 for purchasing $200 worth of goods, what is the rate of his commission?

8. If an agent returns to me $190, after deducting a commission of 5% from the sum received for my goods, how much did he get for the goods?

$190 is what per cent of the sum received?

9. What will be the amount of the commission of an agent who purchases 2 car loads of corn containing 520 bushels each, at 56 cents a bushel, if his commission is $1\frac{3}{4}$%?

229

Review

Change to improper fractions:

1. $4\frac{3}{4}$; $8\frac{2}{6}$; $3\frac{3}{9}$; $9\frac{5}{6}$; $15\frac{2}{7}$; $12\frac{1}{6}$; $13\frac{5}{8}$; $16\frac{2}{6}$.

2. $12\frac{1}{2}$; $16\frac{2}{3}$; $11\frac{1}{9}$; $33\frac{1}{8}$; $66\frac{2}{6}$; $7\frac{11}{20}$; $2\frac{9}{35}$; $4\frac{9}{14}$.

Change to mixed numbers:

3. $\frac{49}{11}$; $\frac{78}{9}$; $\frac{67}{5}$; $\frac{79}{8}$; $\frac{89}{4}$; $\frac{38}{7}$.

4. $\frac{100}{11}$; $\frac{100}{6}$; $\frac{105}{12}$; $\frac{250}{9}$; $\frac{300}{14}$.

Change to smallest terms:

5. $\frac{8}{10}$; $\frac{12}{15}$; $\frac{36}{45}$; $\frac{54}{81}$; $\frac{24}{40}$.

6. $\frac{15}{18}$; $\frac{75}{125}$; $\frac{80}{150}$; $\frac{100}{125}$; $\frac{150}{210}$.

7. Change to 24ths: $\frac{3}{4}$; $\frac{5}{8}$; $\frac{7}{12}$; $\frac{6}{48}$.

8. Change to 36ths: $\frac{5}{6}$; $\frac{2}{9}$; $\frac{7}{18}$; $\frac{14}{72}$.

9. Change to similar fractions: $\frac{3}{5}$; $\frac{5}{6}$; $\frac{7}{15}$.

10. Add $\frac{1}{4}$, $\frac{2}{3}$, $\frac{5}{6}$.   **14.** Multiply $\frac{7}{12}$ by $\frac{6}{7}$.

11. Add $\frac{6}{7}$, $\frac{3}{4}$, $\frac{5}{6}$.   **15.** Multiply $9\frac{1}{4}$ by $8\frac{3}{4}$.

12. From $\frac{9}{8}$ take $\frac{3}{8}$.   **16.** Divide $\frac{7}{8}$ by $\frac{1}{16}$.

13. From $1\frac{1}{16}$ take $\frac{7}{12}$.   **17.** Divide $12\frac{1}{6}$ by $2\frac{1}{6}$.

18. A man sold $\frac{1}{8}$ of his farm at one time and $\frac{1}{6}$ of it at another time. What part had he remaining?

19. What is the value of $4\frac{5}{6}$ cords of wood at \$$5\frac{3}{10}$ a cord?

20. How many house lots, each containing $\frac{1}{8}$ of an acre, can be made from $23\frac{5}{6}$ acres?

21. How many bottles, each containing $\frac{5}{8}$ of a quart, can be filled from $15\frac{3}{4}$ quarts?

22. Copper is $8\frac{4}{6}$ times as heavy as water, and marble $2\frac{4}{6}$ times as heavy as water. Copper is how many times as heavy as marble?

Angles

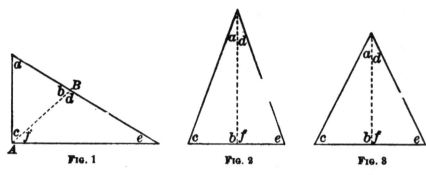

Review pages 78, 123, and 222.

1. How many degrees are there in a right angle?

2. How many degrees are there in the sum of the three angles of any triangle?

3. In the right triangle (Fig. 1), if the angle a is 60°, how large is the angle e?

4. If the line AB divides the right angle at A into 2 equal parts, how large is the angle c?

5. If the angle a is 60°, how large is the angle b?

6. What is the sum of the angles b and d?

7. If e is 30°, how large is the angle d?

8. In the isosceles triangle (Fig. 2), if the whole angle at the vertex is 40°, how large are the angles c and e?

9. If the line from the vertex to the base is perpendicular to the base, how large are the angles b and f_1?

10. What is the sum of all the angles a, b, and c?

11. What is the sum of all the angles d, e, and f_1?

12. If b is equal to f_1 and c is equal to e, how must a and d compare with each other?

13. In the equilateral triangle (Fig. 3) how large is each of the angles?

14. If the line from the vertex to the base is perpendicular to the base, how large are the angles b and f_1?

Miscellaneous Problems

1. What part of a dollar is $12\frac{1}{2}$ cents? $37\frac{1}{2}$ cents? $66\frac{2}{3}$ cents? 80 cents? $87\frac{1}{2}$ cents?

2. What part of a ton is 250 lb.? 400 lb.? 50 lb.? 1600 lb.?

3. If a man earns $\$1\frac{1}{4}$ a day, how much will he earn in the month of September, if there are 4 Sundays in the month?

4. How much is half an acre of land worth at the rate of 8 cents a square foot?

5. If a peck of potatoes costs 30 cents, how many bushels can be bought for $\$6$?

6. At $6\frac{1}{4}$ cents a yard how many yards of cloth can be bought for $\$1$?

7. If a commission of $\frac{1}{8}\%$ is charged for selling stock, what will be the commission for selling stock to the amount of $\$8000$?

8. What will be the commission for selling a car load of potatoes consisting of 640 bushels at 5%, if the potatoes are sold at 43 cents a bushel?

9. A lawyer collected a bill of $\$453.60$, and charged 5% for collecting. How much did he pay to his employer?

10. How many square feet of iron are there in the lateral surface of a cylindrical standpipe which is 28 ft. in diameter and 60 ft. high?

11. How much would it cost to paint both the outer and the inner surface of this standpipe at 9 cents a square yard?

12. When the standpipe is filled with water, what is the pressure per square foot at the bottom?

13. How many square feet are there upon the bottom?

14. What is the total pressure upon the bottom?

232

Commission

1. A commission merchant sold 75 tons of hay at $15.50 a ton. How much was the amount of his commission, reckoned at 3%?

2. A farmer shipped to market 50 boxes of cheese, each containing 42 lb. The commission merchant sold the cheese at 10½ cents a pound, charging 5% for selling. How much money did he return?

3. If $7 is deducted by an agent from $200 received for goods, what is the rate of commission?

4. What is the rate of commission when the agent charges $3 for selling goods to the amount of $150?

5. A farmer shipped to an agent 10 cases of eggs, containing 56 doz. each. They were sold at 23½¢ a dozen. After deducting the freight bill of $3.60, and a commission of 1 cent a dozen for selling, how much did the agent return?

6. Goods were sold by an agent for $800. He deducted his commission and a bill of $15.75 for freight and cartage, and returned $768.25 to the shipper. How much was his commission?

7. What was the rate of his commission?

8. What is the sum of money of which $1.08 is 3%?

9. An agent received $54.30 for selling grain at a commission of 3%. How much did he receive for the grain?

10. If an agent received $30 as his commission for selling grain at 25 cents a bushel on a commission of 2%, how many bushels did he sell?

11. An agent returns to his employer $95 after taking his commission of 5%. What was the whole amount received by the agent?

Pyramids

A pyramid has triangles for its lateral faces, and some kind of a polygon for its base. If the pyramid is regular, the lateral faces are isosceles triangles. The **altitude** of these triangles is the **slant height** of the pyramid.

1. The perimeter of the base of a regular pyramid is 56 in. The slant height is 2 ft. 8 in. Find the lateral surface.

2 ft. 8 in. = 32 in.; 56 × 32 ÷ 2 = 896 sq. in., or 6⅙ sq. ft.

2. Find the lateral surface of a square pyramid whose base is 2 ft. 4 in. square and whose slant height is 8 ft. 9 in.

3. Find the entire surface of a square pyramid whose base is 5 yards square and whose slant height is 34½ feet.

Find the lateral surface of the following pyramids:

4. The base is 9 in. square. The slant height is 16 in.

5. The base is a rectangle 15 in. × 8 in. The slant height is 28 in.

6. The base is an equilateral triangle, each side of which is 3½ ft. The slant height is 6 ft. 4 in.

7. The base is a triangle whose sides are 8 in., 10 in., and 16 in. The slant height is 21 in.

Multiplication of Fractions

To multiply one mixed number by another, in some cases, when the fractions contain only small numbers, it is better to multiply directly without change. In other cases it is easier to change the mixed numbers to improper fractions and multiply together the numerators and the denominators.

See pages 81, 217, and 221.

Find the continued product:

1. $\frac{2}{3} \times \frac{3}{4} \times \frac{4}{5} \times \frac{5}{6}$.

2. $\frac{4}{5} \times \frac{7}{10} \times \frac{5}{6} \times \frac{6}{7}$.

3. $\frac{5}{8} \times \frac{11}{12} \times \frac{10}{7} \times \frac{8}{11}$.

4. $\frac{15}{16} \times \frac{9}{10} \times \frac{3}{5} \times \frac{10}{9}$.

5. $\frac{17}{21}$ of $\frac{18}{25}$ of $\frac{16}{17}$ of $\frac{7}{16}$.

6. $\frac{18}{20}$ of $\frac{22}{33}$ of $\frac{18}{19}$ of $\frac{19}{11}$.

7. $\frac{12}{13}$ of $\frac{15}{17}$ of $\frac{18}{12}$ of $\frac{17}{17}$.

8. $\frac{22}{24}$ of $\frac{15}{16}$ of $\frac{12}{46}$ of $\frac{20}{21}$.

Multiply without change of form:

9. $35\frac{1}{2} \times 24\frac{1}{5}$.

10. $72\frac{2}{3} \times 36\frac{3}{8}$.

11. $45\frac{1}{6} \times 48\frac{1}{5}$.

12. $84\frac{1}{7} \times 28\frac{3}{4}$.

13. $60\frac{1}{12} \times 96\frac{1}{5}$.

14. $55\frac{8}{9} \times 27\frac{1}{11}$.

15. $125\frac{2}{3} \times 30\frac{1}{2}$.

16. $218\frac{1}{4} \times 64\frac{3}{8}$.

17. $328\frac{5}{6} \times 80\frac{3}{4}$.

18. $525\frac{2}{7} \times 84\frac{3}{8}$.

19. $466\frac{2}{3} \times 63\frac{1}{8}$.

20. $875\frac{3}{10} \times 70\frac{1}{5}$.

Multiply by first changing to improper fractions:

21. $5\frac{7}{8} \times 4\frac{2}{3}$.

22. $7\frac{5}{9} \times 9\frac{2}{7}$.

23. $8\frac{3}{5} \times 7\frac{1}{4}$.

24. $9\frac{5}{7} \times 10\frac{1}{11}$.

25. $12\frac{5}{8} \times 9\frac{3}{8}$.

26. $18\frac{2}{5} \times 17\frac{5}{8}$.

27. $104\frac{2}{3} \times 31\frac{3}{7}$.

28. $251\frac{16}{17} \times 42\frac{1}{4}$.

29. $318\frac{13}{14} \times 12\frac{5}{8}$.

30. $236\frac{9}{10} \times 24\frac{5}{8}$.

31. $280\frac{7}{11} \times 57\frac{3}{4}$.

32. $516\frac{1}{18} \times 46\frac{1}{17}$.

Construction

See pages 85, 92, 123, 151, 199, and 218.

1. Construct a right triangle containing an angle of 40°. Measure the sides and find the area.

2. Construct a triangle containing an acute angle and an obtuse angle. Draw the altitude. Measure the base and the altitude and find the area.

3. Construct an isosceles triangle with a base 2 inches long. Draw a dotted line for the altitude. Find the area.

4. Construct an equilateral triangle with a side $2\frac{3}{4}$ inches long. Draw a dotted line for the altitude. Find the area.

5. Construct a parallelogram with an angle of 50° and a side 3 inches long. Draw the diagonals. Which of the angles are equal to each other? Draw the altitude. Find the area.

6. Construct a trapezoid with one of the parallel sides 4 inches long. Draw a dotted line for the altitude. The sum of what angles should equal 180°? Find the area of the trapezoid.

7. Construct a square 3 inches long. Find the center of the square and inscribe a circle. Find the circumference of the circle. Find the area of the circle.

8. Construct a right triangle with the base 2 inches long and the perpendicular $1\frac{1}{2}$ inches long. Measure the hypotenuse. Upon the hypotenuse construct a square. Draw lines dividing this square. Divide the square into half-inch squares. Do the same with the base and the perpendicular. What are the relations between the numbers of squares on the three sides?

236

Original Problems

Make problems and solve them:

1. The ends of a prism are equilateral triangles. Each lateral face of the prism is 12 inches long and 4 inches wide.

2. The perimeter of the end of a triangular prism is 24 inches and the length of the prism 15 inches.

3. The diameter of a roller is 42 inches.

4. It costs $\frac{1}{8}\%$ to send money by express.

5. A bill of $450 is discounted $\frac{1}{2}\%$ for cash.

6. I make 25 cents by selling goods at a profit of $2\frac{1}{2}\%$.

7. By selling wool at a profit of $\frac{1}{2}\%$ a merchant gained $40.

8. An agent sold goods to the amount of $300.

9. An agent's rate of commission is 2%.

10. An agent returns $147.00 to the shipper, for the sale of goods, after deducting a commission of 2%.

11. An agent sold goods to the amount of $200 and kept $6 as his commission.

12. An agent kept $2 as his commission for selling goods to the amount of $40.

13. A man can do $\frac{2}{3}$ of a piece of work in 6 days.

14. A commission of $\frac{1}{8}\%$ is charged for selling $5000 of railroad stock.

15. A cylindrical standpipe is 21 feet in diameter and 50 feet high.

16. An agent deducts $7.50 as his commission for the sale of goods to the amount of $300.

17. 156 is 4% more than some number.

18. 141 is 6% less than a certain number.

Miscellaneous Problems

1. There are 6 ft. in a fathom. How many fathoms are there in half a mile?

2. How many fathoms of cable must be let out to lower the anchor of a ship where the water is 245 ft. deep?

3. If I buy 2¼ acres of land, and cut it up into 9 house lots, how many square rods will each lot contain?

4. Townships in the western states are 6 miles square, and each township is divided into 36 equal square sections. How many acres are there in a farm consisting of ¼ of a section?

5. What is the area of a triangle whose base is 12 ft. 6 in. and whose altitude is 9 ft.?

6. What is the area of a parallelogram if one side is 10⅛ yd. long and the perpendicular distance from this side to the opposite side is 6½ yd.?

7. If a house worth $3000 rents for $500, what per cent does the property yield?

8. A grocer gains 12 cents a gallon by selling molasses at a gain of 25%. What was the cost?

9. A merchant lost $40 by selling a lot of cloth at a loss of 12½%. There were 640 yd. How much did it cost per yard?

10. A house 30 ft. square has a roof in the form of a pyramid. The length of an edge of the roof is 25 feet. Find the slant height.

11. How many square feet are there in the entire roof?

12. What will be the cost of shingles enough to cover the roof at $3.25 per M., if a thousand shingles will cover 110 sq. ft.?

The Barometer

A barometer is an instrument for measuring the pressure of the air. The tube is filled with mercury. The top of the tube is closed so that the air cannot press down upon the mercury there, and the pressure of the air at the bottom holds the mercury up to a certain height in the tube. If the entrance to the tube is 1 sq. in., then the weight of the mercury in the tube will indicate the pressure of the air upon 1 sq. in. If the entrance to the tube is $\frac{1}{4}$ of a sq. in., the weight of the mercury will indicate the pressure of the air upon $\frac{1}{4}$ of a sq. in., etc.

1. If the entrance to the tube of the barometer is 1 sq. in., and the mercury stands 30 in. high in the tube, how many cubic inches of mercury are there?

2. The specific gravity of mercury is 13.59. What is the weight of a cubic foot of mercury?

A cubic foot of water weighs 1000 ounces.

3. Find the weight of a cubic inch of mercury.

4. What will be the weight of the 30 cubic inches of mercury in the tube?

5. When the mercury in the barometer stands 29 in. high, what is the pressure of the atmosphere upon 1 sq. in.?

29 × 7.86 oz. ÷ 16 = how many pounds?

6. During the night preceding a storm the mercury fell from 30 in. to 29.5 in. How much was the pressure of the atmosphere reduced per square inch?

Pressure of the Atmosphere

1. When the mercury in a barometer stands at 30 inches, the pressure of the atmosphere is 14.7 pounds to the square inch. What is the atmospheric pressure when the barometer stands at 27 inches?

In ascending a mountain the pressure of the atmosphere becomes less because the atmosphere above becomes continually less. The barometer falls about 1 inch for each 1000 feet of ascent.

2. A barometer which stood at 29 inches at the foot of a mountain is carried to the top, where it indicates 24½ inches. About what is the pressure at the top?

3. About how much would the barometer fall in ascending to the top of Mt. Washington, which is 6286 feet high?

4. Pikes Peak is 14,147 feet high. About how much difference is there between the height of the barometer at its base and on its summit?

5. When the pressure of the atmosphere is 14.7 pounds per square inch at the base, what is the pressure at the summit?

6. How many sticks, each 1 ft. long, 1 in. wide, and 1 in. thick, will it take to build up a cubic foot?

7. If these sticks were stood up in a single column, one resting upon another endwise, how high would the column extend?

8. What must be the height of a column of air resting upon 1 square inch, to contain a cubic foot?

9. What must be the height of a column of air resting on 1 square inch, to contain 13 cubic feet?

10. 13 cubic feet of air near the surface of the earth will weigh about 1 pound. If the atmosphere at all distances from the earth were equally heavy, how high would it have to extend to press 14.7 pounds upon 1 square inch?

240

Multiplication and Division of Fractions

In multiplying a fraction by 2 we may have either twice as many parts or the same number of parts with the parts twice as large. To multiply a fraction either multiply the numerator or divide the denominator.

$$\tfrac{3}{8} \times 2 = \tfrac{6}{8}, \text{ or } \tfrac{3}{8} \times 2 = \tfrac{3}{4}.$$

In dividing a fraction we may have either one half as many parts or the same number of parts with the parts one half as large. To divide a fraction either divide the numerator or multiply the denominator.

$$\tfrac{4}{5} \div 2 = \tfrac{2}{5}, \text{ or } \tfrac{4}{5} \div 2 = \tfrac{4}{10}.$$

Solve each of the following in two ways:

1. $2 \times \tfrac{1}{6}$
2. $3 \times \tfrac{2}{9}$
3. $4 \times \tfrac{1}{4}$

4. $3 \times \tfrac{5}{6}$
5. $5 \times \tfrac{4}{5}$
6. $4 \times \tfrac{3}{8}$

7. $6 \times \tfrac{1}{12}$
8. $4 \times \tfrac{3}{4}$
9. $8 \times \tfrac{1}{16}$

Solve each of the following in the more convenient way:

10. $\tfrac{1}{6} \times 3$
11. $\tfrac{1}{8} \times 4$
12. $\tfrac{5}{7} \times 5$

13. $\tfrac{5}{6} \times 6$
14. $\tfrac{3}{8} \times 7$
15. $\tfrac{11}{12} \times 4$

16. $\tfrac{7}{11} \times 11$
17. $\tfrac{5}{18} \times 3$
18. $\tfrac{24}{25} \times 4$

Solve each of the following in two ways:

19. $\tfrac{2}{3} \div 2$
20. $\tfrac{4}{5} \div 2$
21. $\tfrac{3}{5} \div 3$

22. $\tfrac{2}{8} \div 2$
23. $\tfrac{9}{10} \div 3$
24. $\tfrac{15}{16} \div 5$

25. $\tfrac{8}{9} \div 4$
26. $\tfrac{14}{15} \div 7$
27. $\tfrac{32}{35} \div 8$

Solve each of the following in the more convenient way:

28. $\tfrac{3}{5} \div 3$
29. $\tfrac{1}{8} \div 4$
30. $\tfrac{8}{9} \div 2$

31. $\tfrac{10}{11} \div 5$
32. $\tfrac{3}{7} \div 4$
33. $\tfrac{5}{8} \div 6$

34. $\tfrac{14}{15} \div 7$
35. $\tfrac{32}{35} \div 16$
36. $\tfrac{17}{39} \div 3$

Miscellaneous Problems

1. A man sold a farm so as to gain $\frac{2}{5}$ of the cost. He gained $1200. What was the cost of the farm?

2. A man sold a house so as to lose $\frac{2}{7}$ of the cost. He received $2500. What was the cost?

3. After $\frac{5}{8}$ of the oil had been sold from a barrel, 24 gal. remained. How many gallons did the barrel contain?

4. A farmer sold $\frac{1}{3}$ and $\frac{2}{5}$ of his crop of wheat, and had 60 bu. remaining. How large was his crop?

5. If a certain number were increased by $\frac{1}{2}$ of itself and $\frac{1}{3}$ of itself, the result would be 55. What is the number?

6. If from a number $\frac{1}{3}$ and $\frac{1}{6}$ of itself are taken, the remainder will be 14. What is the number?

7. If I buy a horse for $100 and sell him for $113, what per cent do I gain?

8. What per cent do I gain by selling a cow for $52, if the cost was $50?

9. What per cent is lost by selling for 23 cents goods which cost 25 cents?

10. If a profit of 20% is gained by selling eggs at 12 cents a dozen, what was the cost?

11. If $12\frac{1}{2}$% is lost by selling cheese at 14 cents a pound, what was the cost?

12. A merchant bought goods for $1000. He sold $\frac{3}{5}$ of them at a gain of 15%, and the remainder at a gain of $12\frac{1}{2}$%. How much did he gain upon the whole?

13. What is the cost of $17\frac{7}{12}$ yd. of cloth at 6 cents a yard?

14. If it costs $32\frac{4}{5}$ to plow 8 acres of land, how much is that an acre?

242

Insurance

Insurance companies agree to pay the owners of property a certain sum if the property should be destroyed by fire, by shipwreck at sea, or by some other accident. In return for this promise the persons insured pay the company each year a sum equal to a certain per cent of the sum which the company promises to pay. The sum paid by the person insured is called the **premium**.

1. How much will it cost each year to have a house insured for $ 2000 at the rate of 2% ?

2. A man insures his house for $ 3400 at $1\frac{1}{4}$%. What is the amount of the premium ?

3. How much will it cost to insure furniture for $ 1500 at $\frac{4}{5}$% ?

4. A merchant's goods are valued at $ 7364. How much will it cost to insure them for $\frac{3}{4}$ of their value at $1\frac{1}{4}$% ?

5. If the property is insured at 2%, and the premium amounts to $ 16, for how much was it insured ?

$ 16 is 2% of what number?

6. The premium for insuring a house at $1\frac{1}{2}$% was $ 30. For what sum was it insured ?

7. A mill valued at $ 27,000 is insured for $\frac{2}{3}$ of its value at $1\frac{2}{3}$%. What is the premium ?

8. How much will it cost to insure a cargo of 4000 tons of coal valued at $ 3.25 per ton for $\frac{3}{4}$ of its value at $\frac{4}{5}$% ?

9. A man paid $ 48 for insuring his shop at 1% on $\frac{3}{4}$ of its value. For what sum was it insured ? What was the value of the shop ?

10. A cargo of 3500 bushels of wheat valued at 60 cents a bushel was insured for $\frac{3}{4}$ of its value at $\frac{7}{8}$%. What was the amount of the premium ?

243

Spheres

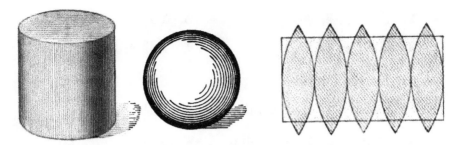

Review page 214.

If the peel of an orange should be taken off in narrow sections, the sections placed in a row, and the ends cut off and filled into the vacant places, the surface of the orange would then correspond to the lateral surface of a cylinder.

The surface of a sphere is equal to the lateral surface of a cylinder of the same height. In other words, the surface of a sphere is equal to the product of its circumference and its diameter.

1. Find the convex surface of a cylinder 20 in. long whose diameter is 20 in.

2. What is the circumference of a sphere whose diameter is 20 in.?

3. Find the surface of a sphere whose diameter is 20 in.

4. Find the entire surface of a cylinder whose diameter is 20 in. and length 20 in.

5. How many square inches are there on the surface of an orange whose diameter is 4.2 in.?

6. Find the number of square feet of spherical surface on a hemisphere whose diameter is 7 ft. 4 in.

7. From a point directly beneath the center of a dome to its outer edge the distance is 15 ft. How many square feet are there in its outer spherical surface?

244

Fractions — Review

1. Add $\frac{2}{5}$, $\frac{2}{3}$, $\frac{4}{15}$, $\frac{5}{6}$.

2. Add $\frac{1}{3}$, $\frac{4}{7}$, $\frac{1}{6}$, $\frac{8}{21}$.

3. $25\frac{1}{4} + 16\frac{5}{9} + 32\frac{1}{6}$.

4. $12\frac{5}{8} + 15\frac{2}{7} + 22\frac{3}{4}$.

5. From $\frac{7}{12}$ take $\frac{1}{8}$.

6. From $35\frac{7}{9}$ take $15\frac{2}{3}$.

7. $52\frac{2}{5} - 17\frac{1}{4}$.

8. $27\frac{1}{2} - 14\frac{3}{11}$.

9. Multiply $275\frac{3}{10}$ by 250.

10. Multiply $24\frac{2}{3}$ by $15\frac{1}{2}$.

11. Multiply $\frac{3}{8}$ by $\frac{1}{2}$.

12. $\frac{3}{4} \times \frac{2}{3}$.

13. $2\frac{1}{2} \times 3\frac{1}{3}$.

14. $648 \times 17\frac{5}{8}$.

15. Divide $\frac{6}{7}$ by 3.

16. Divide $\frac{1}{6}$ by 4.

17. Divide $\frac{3}{4}$ by $\frac{1}{4}$.

18. $\frac{4}{5} \div \frac{1}{3}$.

19. $9\frac{3}{4} \div 3$.

20. $4\frac{1}{3} \div 1\frac{1}{2}$.

21. What is the combined length of 4 boards whose lengths are $5\frac{1}{2}$ ft., $7\frac{3}{10}$ ft., $9\frac{7}{8}$ ft., and $12\frac{4}{5}$ ft. ?

22. If $22\frac{5}{8}$ yards of cloth are taken from a piece containing $45\frac{5}{12}$ yards, how many yards will remain ?

23. If $12\frac{3}{8}$ is $\frac{3}{4}$ of a number, how much is $\frac{1}{2}$ of the number ?

24. $25\frac{1}{8}$ is $\frac{1}{6}$ of a certain number. What is 5 times the number ?

25. If a boy can walk $3\frac{1}{4}$ miles in an hour, how far can he walk in $4\frac{1}{3}$ hours ?

26. A farm of $455\frac{7}{8}$ acres is divided into 7 equal parts. How many acres are there in each part ?

27. How many pieces of cloth $\frac{2}{3}$ of a yard long can be cut from a piece containing $18\frac{3}{4}$ yards ?

28. If a man can walk $12\frac{1}{4}$ rods a minute, how many minutes will it take him to walk a mile ?

Measurements

Review pages 214 and 244.

Find the measurements of a regular prism 3 ft. long:

1. Square prism: width of end 5 in. Lateral surface = ?

2. Square prism: width of end 7 in. Entire surface = ?

3. Triangular prism: width of face 6 in. Lateral surface = ?

4. Triangular prism: perimeter of end 8 in. Lateral surface = ?

Find the measurements of a cylinder 2 ft. long:

5. The circumference, if the diameter is 15 inches.

6. Surface of ends, if the diameter is 15 inches.

7. The entire surface, if the diameter is 15 inches.

8. The circumference, if the radius is 12 inches.

9. The lateral surface, if the diameter is 9 inches.

10. The entire surface, if the diameter is 10 inches.

11. The lateral surface, if the diameter is 2 feet.

12. The entire surface, if the radius is $1\frac{1}{4}$ feet.

Find the measurements of a sphere:

13. Circumference, if the diameter is 10 feet.

14. Surface, if the diameter is 10 feet.

15. Diameter, if the circumference is 32 feet.

16. Surface, if the circumference is 32 feet.

17. Circumference, if the radius is 4 feet.

18. Surface, if the radius is 7 feet.

19. Radius, if the circumference is 20 feet.

20. Surface, if the circumference is 40 feet.

Original Problems

Make problems and solve them:

1. When a certain number is increased by $\frac{1}{2}$ and $\frac{1}{4}$ of itself the result is 35.

2. I bought a horse for $75 and sold him for $90.

3. I sold for 55 cents goods which cost 60 cents.

4. A barrel is 30 inches long, and its average diameter is 20 inches.

5. The diameter of a sphere is 4 ft. 8 in.

6. The distance around a hemispherical dome is 180 feet.

7. From a piece of cloth containing $47\frac{3}{8}$ yards a part has been sold.

8. A man has his house insured for $3000 at $1\frac{1}{2}$%.

9. Property valued at $6000 is insured for $\frac{2}{3}$ of its value.

10. A man paid $60 for insuring property at 2%.

11. The diameter of a cylinder is 7 inches, and its length 10 inches.

12. A yard stick was 2 inches too short. A piece of cloth was measured with it.

13. The nights are 1 hr. 30 min. longer than the days.

14. Goods which cost 60 cents are sold for 48 cents.

15. Goods which cost 25 cents are sold for 29 cents.

16. $13\frac{2}{3}$ yards of cloth have been taken from a large piece.

17. A man can walk $3\frac{5}{8}$ miles in an hour.

18. The diameter of a cylinder is 1 ft. 10 in.

19. The diameter of a globe is 15 inches.

20. The polar diameter of the earth is 7900 miles.

247

Tables

LIQUID MEASURE

4 gills (gi.)	= 1 pint	(pt.).
2 pints	= 1 quart	(qt.).
4 quarts	= 1 gallon	(gal.).
1 gal.	= 231 cubic inches.	

DRY MEASURE

2 pints (pt.)	= 1 quart	(qt.).
8 quarts	= 1 peck	(pk.).
4 pecks	= 1 bushel (bu.).	
1 bushel	= 2150.42 cubic inches.	

LINEAR MEASURE

12 inches (in.)	= 1 foot	(ft.).
3 feet	= 1 yard	(yd.).
5½ yards	= 1 rod	(rd.).
16½ feet	= 1 rod.	
320 rods	= 1 mile	(mi.).
1760 yards	= 1 mile.	
5280 feet	= 1 mile.	

SURFACE MEASURE

144 square inches (sq. in.)	= 1 square foot	(sq. ft.).
9 square feet	= 1 square yard	(sq. yd.).
30¼ square yards	= 1 square rod	(sq. rd.).
272¼ square feet	= 1 square rod.	
160 square rods	= 1 acre	(A.).
4840 square yards	= 1 acre.	
43560 square feet	= 1 acre.	
640 acres	= 1 square mile	(sq. mi.).

SURVEYORS' MEASURE

Long Measure

7.92 in.	= 1 link	(l.).
100 links	= 1 chain	(ch.).
80 chains	= 1 mile	(mi.).

Square Measure

16 sq. rd.	= 1 sq. ch.
10 sq. ch.	= 1 A.
640 A.	= 1 sq. mi.
1 sq. mi.	= 1 section (sec.).
36 sec.	= 1 township.

Tables

ANGULAR MEASURE

60 seconds (″) = 1 minute (′).
60 minutes   = 1 degree (°).
360 degrees   = 1 circle.

SOLID OR CUBIC MEASURE

1728 cubic inches (cu. in.) = 1 cubic foot (cu. ft.).
27 cubic feet   = 1 cubic yard (cu. yd.).

WOOD MEASURE

16 cubic feet   = 1 cord foot (cd. ft.).
8 cord feet, or
128 cubic feet   } = 1 cord   (cd.).

AVOIRDUPOIS WEIGHT

Avoirdupois Weight is used in weighing all articles except gold, silver, and precious stones.

16 ounces (oz.) = 1 pound   (lb.).
100 pounds   = 1 hundredweight (cwt.).
2000 pounds   = 1 ton   (T.).

The long ton is used in the United States customhouses and in wholesale transactions in iron and coal.

112 pounds Avoirdupois = 1 long hundredweight.
2240 pounds Avoirdupois = 1 long ton.

The pound Avoirdupois contains 7000 grains.

TROY WEIGHT

Troy Weight is used in weighing gold, silver, and jewels.

24 grains (gr.)   = 1 pennyweight (pwt.).
20 pennyweights = 1 ounce   (oz.).
12 ounces   = 1 pound   (lb.).

Tables

Apothecaries' Weight is used in mixing medicines.

20 grains (gr.) = 1 scruple (Э).	8 drams = 1 ounce (℥).
3 scruples = 1 dram (℈).	12 ounces = 1 pound (℔).

The pound Troy and the Apothecaries' pound are equal, each weighing 5760 grains. The pound Avoirdupois weighs 7000 Troy or Apothecary grains.

The ounce Troy and the Apothecaries' ounce are each 480 grains; the ounce Avoirdupois is 437½ grains.

Time Measure

60 seconds (sec.) = 1 minute (min.).		365 days = 1 common
60 minutes = 1 hour (hr.).		year (yr.).
24 hours = 1 day (da.).		366 days = 1 leap year.
7 days = 1 week (wk.).		100 years = 1 century.

The names of the months (mo.), called calendar months, and the number of days in each are:

	da.		da.
1. January (Jan.) 31		7. July 31	
2. February (Feb.) . 28 or 29		8. August (Aug.) . . . 31	
3. March (Mar.) 31		9. September (Sept.) . . 30	
4. April (Apr.) 30		10. October (Oct.) 31	
5. May 31		11. November (Nov.) . . 30	
6. June 30		12. December (Dec.) . . . 31	

United States Money	English Money
10 mills (m.) = 1 cent (¢).	4 farthings (far.) = 1 penny (d.).
10 cents = 1 dime (d.).	12 pence = 1 shilling (sh.).
10 dimes = 1 dollar ($).	20 shillings = 1 pound or
10 dollars = 1 Eagle (E.).	sovereign (£).

Miscellaneous Table

12 units = 1 dozen.	20 units = 1 score.
12 dozen = 1 gross.	24 sheets = 1 quire.
12 gross = 1 great gross.	20 quires = 1 ream.

For Reference

Acute angle. An acute angle is smaller than a right angle.

Altitude. The altitude of a figure is the perpendicular distance from the base to the highest point, or the point farthest from the base.

Angle. An angle is the difference in direction of two lines.

Area. The area of a square, a rectangle, or a parallelogram, is equal to the product of its base and its altitude.

The area of a triangle is equal to one half of the product of its base and altitude.

The area of a trapezoid is equal to the product of the average between its two parallel sides and the perpendicular distance between them.

The area of a circle is equal to one half of the product of its radius and circumference.

Atmospheric pressure. The pressure of the atmosphere upon each square inch at the level of the sea is about 14.7 pounds.

Barometer. The barometer falls about an inch in each 1000 feet of ascent.

Base. The base of a figure is the line or plane upon which it is supposed to rest.

Bricks. A brick is 8 in. long, 4 in. wide, and 2 in. thick. On the side of a wall 7 bricks with the mortar cover about 1 sq. ft. About 22 bricks with the mortar will fill 1 cu. ft.

Bushels. A bushel contains 2150.42 cu. in. There are about $\frac{4}{5}$ as many bushels in a bin as there are cubic feet. Of fruits and vegetables there are about $\frac{2}{3}$ as many bushels as cubic feet.

Circle. A circle is a plane figure bounded by a curved line, all points of which are at equal distances from the center.

Circumference. The circumference of a circle is its boundary line. The circumference of a circle is 3.1416 times its diameter.

Clapboards. A clapboard is usually 4 ft. long and 6 in. wide. There are 25 in a bunch. A clapboard will cover about one square foot of surface.

Cone. A cone may be regarded as a pyramid whose base contains an infinitely large number of sides, so as to become a circle.

For Reference

Cylinder. A cylinder is a solid whose ends or bases are circles.

Diagonal. A diagonal is a straight line connecting two angles that are not adjacent to each other.

Diameter. The diameter is the distance across the circle through the center.

Equilateral triangle. An equilateral triangle is a triangle all of whose sides are equal.

Falling bodies. A heavy body will fall about 16 feet in one second, $2 \times 2 \times 16$ feet in two seconds, $3 \times 3 \times 16$ feet in three seconds, etc. The distance through which a body falls is proportional to the square of the number representing the time of the fall.

Gallons. A gallon contains 231 cu. in. A gallon of water weighs about $8\frac{1}{4}$ lb. A cubic foot of water weighs $62\frac{1}{2}$ lb.

Hexagon. A hexagon is a polygon having six sides.

Isosceles triangle. An isosceles triangle is a triangle two of whose sides are equal.

Lateral surface. The lateral surface of a solid is the surface of its sides exclusive of its ends or bases.

The lateral surface of a cylinder is equal to the circumference of its base multiplied by its altitude.

The lateral surface of a regular pyramid or cone is equal to one half of the product of the perimeter or circumference of the base and the slant height.

Laths. A lath is 4 ft. long and $1\frac{1}{2}$ in. wide. Laths are usually left $\frac{3}{8}$ of an inch apart when nailed. There are 50 laths in a bunch. A bunch will cover 3 square yards of surface.

Light. Light travels at the rate of about 186,000 miles a second.

Oblong. An oblong is a rectangle whose length exceeds its width.

Obtuse angle. An obtuse angle is larger than a right angle.

Paper. In America wall paper is usually 18 in. wide. A single roll is 24 ft. long, a double roll 48 ft. long.

Parallelogram. A parallelogram is a quadrilateral whose opposite sides are parallel to each other.

Pentagon. A pentagon is a polygon having five sides.

For Reference

Perimeter. The perimeter of a plane figure is its boundary line.

Plane. A plane is a surface such that a straight line joining any two points in it lies wholly within that surface.

Polygon. A polygon is a plane figure bounded by straight lines.
In a regular polygon the sides are all equal and the angles are all equal.

Pressure. The pressure of liquids at any point is equal in all directions.

Prism. A prism is a solid whose ends are polygons and whose sides are parallelograms.

Pyramid. A pyramid is a solid whose base is a polygon and whose lateral faces are triangles.

Quadrilateral. A quadrilateral is a polygon having four sides.

Radius. The radius is the distance from the center of a circle to the circumference.

Rectangle. A rectangle is a quadrilateral all of whose angles are right angles.

Right angle. A right angle is an angle of 90°.

Right triangle. A right triangle is a triangle which has one right angle.

Shingles. Shingles are packed and bound in bunches. Four bunches of shingles make 1000. A thousand shingles will cover 100 square feet, when laid with 4 inches exposed to the weather.

Slant height. The slant height of a pyramid is the perpendicular distance from its vertex to one of the sides of its base.

Sound. Sound travels in the air at the rate of about 1100 feet a second, and in the water at the rate of about 4700 feet a second.

Specific gravity. The specific gravity of a substance is the ratio of its weight to the weight of water.

Sphere. A sphere is a solid, all points of whose surface are equally distant from the center.

Square. A square is a quadrilateral, all of whose angles are right angles and all of whose sides are equal.

Stone masonry. Stone is reckoned in cubic feet, in cords, or in perches. A perch is 24¾ cu. ft.

For Reference

Surface. The surface of a sphere is equal to the product of its diameter and its circumference.

See also lateral surface.

Trapezoid. A trapezoid is a quadrilateral, two of whose sides are parallel to each other.

Triangle. A triangle is a polygon having three sides.

The sum of the angles of any triangle equals 180°.

Vertex. The vertex of a figure is its highest point, or the point of the angle opposite the base.

Volume. The volume of a solid is the number of cubic units which it contains.

The volume of a prism or of a cylinder is equal to the area of one end multiplied by the length.

The volume of a pyramid or of a cone is equal to one third of the product of the area of the base and the altitude.

The volume of a sphere is equal to its surface multiplied by one third of its radius.

Water. A cubic foot of water weighs 1000 ounces.

Water in freezing expands about 7½ per cent.

Protractors

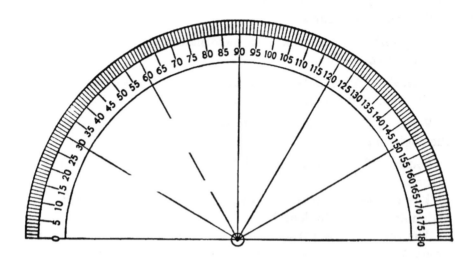

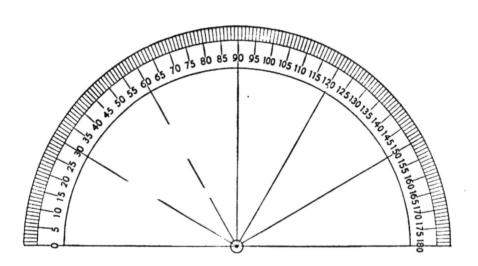

Paste one of these upon a piece of stiff paper or cardboard and cut it out to use with pp. 82, 88, etc.

ANSWERS

Page 10. — 1. 36,107.   2. 30,800.   3. 61,135.   4. 51,137.
5. 3,497,978.   6. 92,179,429.   7. 261,726,487.

Page 11. — 1. 1,738,627.   2. $771,000.   3. 14,855.   4. 749.   5. 10,712.
6. 38,400.   7. 43,384.   8. About 343.   9. 755,040.   10. 2112.   11. 41,060.

Page 12. — 1. 9129.   2. 8191.   3. 6427.   4. 5301.   5. 2862.   6. 1200.
7. 2322.   8. 5189.   9. 4447.   10. 2061.   11. 2798.   12. 3542.   13. 2999.
14. 6991.   15. 4999.   16. 999.   17. 105,300.   18. 107,172.   19. 158,175.
20. 214,312.   21. 115,200.   22. 207,360.   23. 314,900.   24. 447,670.
25. 1,390,592.   26. 5,947,130.   27. 2,994,780.   28. 5,423,000.
29. 2,442,000.   30. 3,480,000.   31. 3,236,100.   32. 6,300,000.   33. 86.
34. 32.   35. 56.   36. 18.   37. 32.   38. 60.   39. 26.   40. 78.   41. 48.
42. 69.   43. 150.   44. 33.   45. 8.   46. 22.   47. 6.   48. 4.

Page 17. — 2. 13.   3. 17.   4. 32.   5. 56.   6. 40¢.   7. 30¢.   8. ⅛.
9. ¼.   10. 2¢.   11. 10 days.   12. 12¢.   13. 48¢.   14. 15.   15. 6.   16. 5.

Page 21. — 1. $7.14.   2. $49.50.   3. $56.70.   4. $12.71.   5. $522.50.
6. $780.   7. $229.13.   8. $864.   9. $101.25.   10. $918.   11. $5.86.
12. $21.   13. $55.08.   14. $1190.   15. $2500.   16. $52.   17. $74.25.
18. $52.50.   19. $4.75.   20. $9.28.   21. $66.   22. $60.40.   23. $23.75.
24. $132.   25. $601.25.

Page 26. — 3. 10 gal.   4. 12 gal. 1 qt. 1 pt.   5. 16 bu. 3 pk. 3 qt.
6. 27 bu. 2 pk. 6 qt.   7. 6 bu. 1 pk. 3 qt.   8. 1 yd. 1 ft. 2 in.   9. 13 bu.
3 pk. 3 qt.   10. 1 yd. 2 ft. 11 in.

Page 27. — 1. 32.   2. 528 ft.   3. 1584.   4. 99.   5. $11.47.   6. 176.
7. 10.   8. 1760.   9. 5280.   10. 20.   11. 754⅘.   12. 165 rd.

Page 28. — 1. 200 pt.   2. 23 pk.   3. 76 in.   4. 147 pt.   5. 560 qt.
6. 85 ft.   7. 57 gal.   8. 20 bu.   9. 21 yd.   10. 80 gal. 3 qt.   11. 43 bu.
3 pk.   12. 83 yd. 1 ft.   13. 62 pt.   14. 648 qt.   15. 461 in.   16. 87 qt. 1 pt.
17. 33 pk. 5 qt.   18. 8 ft. 4 in.   19. 135 pt.   20. 191 qt.   21. 20 yd. 2 ft.
6 in.   22. 18 gal. 3 qt. 1 pt.   23. 977 qt.   24. 644 in.   25. 17 yd. 2 ft. 4 in.

Page 32. — 1. 12 gal. 3 qt.   2. 32 bu. 3 qt.   3. 56 yd. 2 in.   4. 2 gal.
3 qt.   5. 3 bu. 2 pk. 3 qt.   6. 11 yd. 11 in.   7. 10 gal. 3 qt.   8. 16 gal.
2 qt. 1 pt.   9. 29 bu. 2 pk. 7 qt.   10. 24 yd. 1 ft. 7 in.   11. 82 yd. 1 ft.
12. 21 yd. 6 in.

Answers

Page 35. — **1.** 109. **2.** 114. **3.** $40\frac{7}{8}$. **4.** Nearly 7. **5.** 121. **6.** About 15. **7.** $\$970$. **8.** $43\frac{3}{4}$ da. **9.** 9458 ft.

Page 36. — **3.** 135. **4.** $2\frac{3}{4}$. **5.** 4500 lb. **6.** 3. **7.** $22\frac{2}{3}$. **8.** $\$6.25$. **9.** $\$.20$. **10.** 160. **11.** $\$.24$. **12.** 224.

Page 37. — **18.** $15\frac{1}{8}$. **19.** $22\frac{3}{8}$. **20.** $32\frac{1}{6}$. **21.** $44\frac{1}{8}$. **22.** $68\frac{1}{2}$. **23.** $5\frac{2}{3}$. **24.** $3\frac{7}{9}$. **25.** $10\frac{1}{6}$. **26.** $42\frac{5}{12}$. **27.** $28\frac{11}{12}$.

Page 38. — **1.** $10\frac{5}{9}$. **2.** $23\frac{5}{9}$. **3.** 34. **4.** 82. **5.** $198\frac{2}{3}$. **6.** $303\frac{3}{4}$. **7.** $592\frac{1}{12}$. **8.** $982\frac{1}{2}$. **9.** $4\frac{1}{9}$. **10.** $18\frac{1}{3}$. **11.** 4. **12.** 3. **13.** $3\frac{1}{12}$. **14.** 3. **15.** $1\frac{1}{2}$. **16.** $1\frac{1}{4}$. **17.** $60\frac{6}{7}$. **18.** $23\frac{11}{12}$. **19.** $36\frac{7}{8}$ ¢. **20.** $82\frac{4}{5}$ ¢. **21.** $\$5\frac{3}{5}$. **22.** $98\frac{8}{9}$ ¢. **23.** 71. **24.** 45. **25.** 85. **26.** $78\frac{2}{3}$ ¢. **27.** $97\frac{1}{2}$ ¢. **28.** $\$4.60$. **29.** $\$7\frac{1}{2}$.

Page 41. — **6.** 432. **7.** 1296. **8.** 47. **9.** 3322. **10.** 3. **11.** $6\frac{2}{3}$. **12.** 4. **13.** 9. **14.** 16. **15.** 25.

Page 42. — **3.** 11 gal. 1 qt. **4.** 31 bu. 2 pk. **5.** 53 bu. 2 pk. 3 qt. **6.** 39 yd. 1 ft. 8 in. **7.** 4 bu. 1 pk. 2 qt. **8.** 2 bu. 1 pk. 3 qt. **9.** 3 yd. 1 ft. 4 in. **10.** 1 yd. 2 ft. 3 in. **11.** 102 gal. **12.** 113 bu. 1 pk. 2 qt. **13.** 2 bu. 4 pk. 4 qt. **14.** 3 yd. $10\frac{2}{3}$ in. **15.** 385 ft. 6 in. **16.** 21 bu. 1 pk. $5\frac{1}{4}$ qt.

Page 44. — **1.** 24. **2.** 12. **3.** 39 ¢. **4.** 32 da. **5.** 80 ¢. **6.** 80 ¢. **7.** $\$1$. **8.** $\$3$. **9.** $\$3.60$. **10.** 26. **11.** 200. **12.** $\$8.25$. **13.** 2640.

Page 45. — **1.** 24 sq. in. **2.** 84 sq. ft. **3.** $\frac{1}{2}$. **4.** 36 ft. **5.** 75. **6.** 98. **7.** 108. **8.** 2048. **9.** 33. **10.** 50. **11.** $\$9$.

Page 46. — **1.** $\frac{8}{9}$. **2.** $1\frac{1}{9}$. **3.** $1\frac{1}{2}$. **4.** $1\frac{1}{6}$. **5.** $1\frac{1}{2}$. **6.** $1\frac{7}{8}$. **7.** $1\frac{5}{9}$. **8.** $1\frac{2}{3}$. **9.** $1\frac{7}{8}$. **10.** $14\frac{1}{2}$. **11.** $13\frac{1}{8}$. **12.** $22\frac{2}{9}$. **13.** $\frac{2}{3}$. **14.** $\frac{7}{9}$. **15.** $\frac{3}{10}$. **16.** $1\frac{1}{10}$. **17.** $\frac{1}{2}$. **18.** $\frac{3}{8}$. **19.** $3\frac{2}{9}$. **20.** $3\frac{4}{9}$. **21.** $7\frac{1}{8}$. **22.** $11\frac{7}{10}$. **23.** $5\frac{5}{6}$. **24.** $\frac{9}{10}$. **25.** 6. **26.** $3\frac{1}{3}$. **27.** $6\frac{2}{3}$. **28.** $6\frac{3}{10}$. **29.** $4\frac{3}{8}$. **30.** $7\frac{1}{2}$. **31.** $15\frac{1}{3}$. **32.** $31\frac{1}{5}$. **33.** $16\frac{8}{9}$. **34.** $25\frac{4}{5}$. **35.** 46. **36.** 85. **37.** $\frac{4}{9}$. **38.** $\frac{3}{10}$. **39.** 2. **40.** 8. **41.** 2. **42.** $2\frac{2}{3}$. **43.** $2\frac{1}{9}$. **44.** $2\frac{1}{6}$. **45.** $4\frac{3}{10}$. **46.** $5\frac{3}{10}$. **47.** 3. **48.** 3.

Page 48. — **1.** 25. **2.** 64. **3.** 60. **4.** 8. **5.** 300. **6.** 5400. **7.** $170\frac{2}{3}$. **8.** 968. **9.** About $5\frac{2}{3}$. **10.** About 23. **11.** 200. **12.** 150. **13.** $33\frac{1}{3}$. **14.** 700.

Page 49. — **1.** $223\frac{7}{8}$. **2.** $158\frac{1}{8}$. **3.** $194\frac{8}{9}$. **4.** $201\frac{4}{5}$. **5.** $210\frac{5}{12}$. **6.** $177\frac{3}{8}$. **7.** $1021\frac{1}{8}$. **8.** $1229\frac{1}{2}$. **9.** $1086\frac{1}{2}$. **10.** $1039\frac{2}{9}$. **11.** $891\frac{7}{12}$. **12.** $938\frac{1}{6}$. **13.** $42\frac{1}{4}$. **14.** $56\frac{2}{3}$. **15.** $19\frac{2}{3}$. **16.** $24\frac{1}{2}$. **17.** $39\frac{5}{12}$. **18.** $9\frac{2}{3}$. **19.** $123\frac{3}{8}$. **20.** $116\frac{3}{10}$. **21.** $196\frac{5}{6}$. **22.** $292\frac{5}{12}$. **23.** $127\frac{3}{8}$. **24.** $707\frac{1}{3}$. **25.** 6612. **26.** 14,510. **27.** 29,029. **28.** 23,030. **29.** 64,197. **30.** 17,201. **31.** 6797. **32.** 29,552. **33.** 53,680. **34.** 12,013. **35.** 24,432. **36.** 44,156. **37.** $82\frac{1}{4}$. **38.** $62\frac{1}{5}$. **39.** $150\frac{2}{11}$. **40.** $157\frac{3}{10}$. **41.** $97\frac{1}{7}$. **42.** $197\frac{5}{12}$. **43.** $206\frac{2}{9}$. **44.** $151\frac{1}{8}$. **45.** $123\frac{3}{10}$. **46.** $61\frac{1}{8}$. **47.** $66\frac{1}{7}$. **48.** $170\frac{2}{11}$.

Page 50. — **2.** 2178. **3.** 65. **4.** 5445. **5.** $\$5308.88$. **6.** $\$20,092.05$. **7.** 25. **12.** 2040. **13.** 16. **14.** 113 A. 127 sq. rd. **15.** 116 A. 30 sq. rd.

Answers

Page 51. — **1.** \$5566.06. **2.** \$6335.84. **3.** \$31,617.55. **4.** \$43,096.49. **5.** \$2.86. **6.** \$94.30. **7.** \$202.50. **8.** \$45.56. **9.** \$4.60. **10.** \$5.72.

Page 52. — **1.** 29 gal. **2.** 288 gal. 3 qt. **3.** 218 bu. 20 qt. **4.** 214 yd. 9 in. **5.** 4 gal. 1 qt. $1\frac{3}{4}$ pt. **6.** 6 bu. 2 pk. $1\frac{5}{7}$ pt. **7.** 12 bu. 1 pk. 6 qt. $1\frac{1}{3}$ pt. **8.** 4 ft. $8\frac{3}{4}$ in. **9.** 780 bu. 15 qt. **10.** 1033 ft. 4 in. **11.** 6 gal. $2\frac{3}{10}$ qt. **12.** 3 yd. 1 ft. $8\frac{3}{4}$ in. **13.** 46 gal. 2 qt. **14.** 46 bu. 3 pk. **15.** 45 bu. 2 pk. 7 qt. **16.** 183. **17.** 15 yd. 1 ft. 8 in. **18.** 8 ft. 3 in. **19.** 2 bu. 2 pk. 7 qt.

Page 53. — **10.** $3\frac{1}{2}$. **11.** 192. **12.** 288. **13.** \$5.25. **14.** \$28. **15.** 96. **16.** $10\frac{1}{2}$ tons.

Page 55. — **1.** 100. **2.** 169. **3.** 225. **4.** 400. **5.** 900. **6.** 1600. **7.** 90. **8.** 198. **9.** 336. **10.** 528. **11.** 1102. **12.** 1584. **13.** 9. **14.** 36. **15.** 144. **16.** 144. **17.** 9. **18.** 6. **19.** 240. **20.** 126. **21.** 10. **22.** 20.

Page 56. — **1.** 4. **2.** $40\frac{5}{16}$. **3.** 640. **4.** 40. **5.** 40. **6.** 33. **7.** 7680. **8.** $5\frac{1}{2}$ mi. **9.** 4. **10.** 5000. **11.** About 39. **12.** About 12. **13.** About 473.

Page 57. — **1.** 25¢. **2.** 9¢. **3.** 74¢. **4.** \$3.16. **5.** \$187.85. **6.** \$1180.39. **7.** \$224.52. **8.** \$13.50.

Page 58. — **1.** 14 gal. 2 qt. **2.** 3 bu. 3 pk. 2 qt. **3.** 54 bu. 2 pk. 6 qt. **4.** 37. **5.** 5. **6.** 86. **7.** \$301. **8.** \$101.79. **9.** 400. **10.** $1\frac{1}{2}$. **11.** 2250. **12.** 7. **13.** 34.

Page 59. — **1.** 64. **2.** 144. **3.** 324. **4.** 576. **5.** 1296. **6.** 7056. **7.** 45. **8.** 80. **9.** 270. **10.** 384. **11.** 432. **12.** 1728. **13.** 27 gal. 3 qt. **14.** 13 bu. 1 pk. 2 qt. **15.** 26 yd. 2 ft. 5 in. **16.** 7 gal. 2 qt. **17.** 10 bu. 1 pk. 3 qt. **18.** 14 yd. 1 ft. 7 in. **19.** 64 gal. 1 qt. 1 pt. **20.** 607 bu. 1 pk. 4 qt. **21.** 436 yd. 2 ft. **22.** 3 gal. 1 qt. $\frac{1}{3}$ pt. **23.** 3 bu. 1 pk. 4 qt. **24.** 9 yd. 1 ft. $6\frac{1}{9}$ in.

Page 62. — **2.** 640. **3.** 160. **4.** 51,200. **5.** 160. **6.** 640. **7.** \$384. **8.** 2640. **9.** $10\frac{14}{25}$ min.

Page 63. — **15.** $29\frac{7}{15}$. **16.** $32\frac{8}{15}$. **17.** $22\frac{5}{7}$. **18.** $193\frac{4}{7}$. **19.** $324\frac{1}{3}$. **20.** $5\frac{1}{3}$. **21.** $5\frac{4}{15}$. **22.** $9\frac{1}{14}$. **23.** $12\frac{3}{14}$. **24.** $25\frac{11}{15}$.

Page 64. — **1.** \$6.88. **2.** \$11. **3.** \$9.80. **4.** $2\frac{1}{2}$. **5.** \$2.38. **6.** \$2.40. **7.** \$1.40. **8.** $26\frac{7}{20}$. **9.** \$48.33.

Page 65. — **1.** $104\frac{7}{8}$. **2.** $162\frac{5}{9}$. **3.** 193. **4.** $193\frac{1}{4}$. **5.** $200\frac{4}{5}$. **6.** $166\frac{1}{14}$. **7.** $235\frac{11}{15}$. **8.** $183\frac{1}{12}$. **9.** $1002\frac{3}{8}$. **10.** $1422\frac{2}{5}$. **11.** $1267\frac{1}{2}$. **12.** $1046\frac{3}{14}$. **13.** $1540\frac{5}{12}$. **14.** $875\frac{1}{2}$. **15.** $1203\frac{2}{15}$. **16.** $1265\frac{2}{9}$. **17.** $19\frac{1}{8}$. **18.** $13\frac{5}{12}$. **19.** $68\frac{1}{6}$. **20.** $25\frac{3}{14}$. **21.** $47\frac{1}{12}$. **22.** $40\frac{1}{5}$. **23.** $17\frac{7}{8}$. **24.** $39\frac{5}{8}$. **25.** $37\frac{1}{2}$. **26.** $206\frac{1}{6}$. **27.** $704\frac{5}{7}$. **28.** $527\frac{1}{2}$. **29.** $185\frac{4}{15}$. **30.** $599\frac{11}{12}$. **31.** $197\frac{1}{9}$. **32.** $309\frac{13}{14}$. **33.** 385. **34.** 930. **35.** 4020. **36.** 3776. **37.** 2985. **38.** 3447. **39.** 7096. **40.** 3324. **41.** 27150. **42.** 95,050. **43.** 202,620. **44.** 69,360. **45.** 467,038. **46.** 152,000. **47.** 446,519. **48.** 221,278.

Answers

Page 66. — 1. 3. **2.** 20. **3.** 18. **4.** 2. **5.** 16. **6.** 120. **7.** 20. **8.** 15. **9.** 50. **10.** 540. **11.** 328. **12.** 1300.

Page 67. — 1. 34. **2.** 49$\frac{1}{2}$ ft. **3.** 38$\frac{1}{4}$ yd. **4.** No difference. **5.** 71$\frac{1}{4}$ **6.** 68. **7.** 21. **8.** 280$\frac{1}{4}$. **9.** 1122.

Page 68. — 2. 5. **3.** 8. **4.** 9. **5.** 9. **6.** 320$\frac{1}{2}$. **7.** 38$\frac{3}{4}$. **8.** 2$\frac{1}{4}$. **9.** 4. **10.** Total, 5$\frac{1}{2}$ in. ; average, 1$\frac{1}{10}$ in.

Page 69. — 1. 105. **2.** 4800. **3.** 2 da. 2 hr. **4.** 26 hr. 18 min. 10 sec. **5.** 64 hr. 33 min. **6.** 3 hr. 19 min. 15 sec. **7.** 300 mi. **8.** 5 hr. **9.** $\frac{1}{4}$. **10.** 3010. **11.** 20 mi. **12.** 55 min. **13.** 4 hr. 20 min. **14.** $730.60.

Page 70. — 5. 64. **6.** 152. · **7.** 6. **8.** 8. **9.** 2 mo. 4 da. **10.** 4 mo. 3 da. **11.** 4 mo. 11 da. **12.** 2 mo. 21 da. **13.** 1 mo. 18 da. **14.** 3 mo. **15.** 2 mo. 19 da. **16.** 2 mo. 16 da. **17.** 2 mo. 20 da. **18.** 2 mo. 15 da. **19.** 4 mo. 21 da. **20.** 2 mo. 22 da.

Page 72. — 1. 1$\frac{1}{4}$. **2.** 1$\frac{5}{14}$. **3.** 1$\frac{1}{2}$. **4.** 1$\frac{1}{15}$. **5.** 1$\frac{4}{13}$. **6.** 1$\frac{1}{2}$. **7.** $\frac{6}{7}$. **8.** 2. **9.** 1$\frac{2}{15}$. **10.** 1$\frac{3}{4}$. **11.** 10. **12.** 20. **13.** 12. **14.** 6$\frac{1}{4}$. **15.** 5. **16.** 12. **17.** 18. **18.** 15. **19.** 8. **20.** 13. **21.** 18. **22.** 51$\frac{3}{4}$. **23.** 129$\frac{1}{4}$. **24.** 310$\frac{1}{4}$. **25.** 1$\frac{1}{4}$. **26.** 1$\frac{1}{10}$. **27.** 13 qt. **28.** 14 bu. 4 qt. **29.** 3$\frac{1}{4}$ da. **30.** 16 yd. 3 in. **31.** 1 mo. 11 da. **32.** 2 mo. **33.** 2 mo. 11 da. **34.** 1 mo. 2 da. **35.** 1 mo. 29 da. **36.** 30 da. **37.** 1 mo. 4 da. **38.** 1 mo. **39.** 2 mo. **40.** 1 mo. 8 da.

Page 74. — 1. 3$\frac{1}{2}$ ft. **2.** 10 rd. **3.** 1,312,000 A. **4.** About 7. **5.** 390. **6.** 3$\frac{3}{10}$. **7.** $92.40. **8.** 12 yr. 4 mo. **9.** 12°. **10.** 56 yr. 2 mo. 3 da. **11.** 311. **12.** 40°. **13.** 36°.

Page 75. — 1. 70 pt. **2.** 797 qt. **3.** 584 in. **4.** 952 in. **5.** 3923 sq. in. · **6.** 1461$\frac{1}{4}$ sq. ft. **7.** 8 gal. 1 qt. 1 pt. **8.** 2 bu. 1 pk. 3 qt. **9.** 5 yd. 2 ft. 3 in. **10.** 1 rd. 1 ft. **11.** 2 sq. ft. 12 sq. in. **12.** 1 sq. rd. 47$\frac{3}{4}$ sq. ft. **13.** 22 gal. 1 qt. 1 pt. **14.** 83 bu. 2 pk. 4 qt. **15.** 39 yd. 1 ft. 4 in. **16.** 30 rd. 14 ft. 1 in. **17.** 4 sq. rd. 54 sq. ft. 120 sq. in. **18.** 381 A. 90 sq. rd. **19.** 12 bu. 2 pk. 7 qt. **20.** 13 yd. 10 in. **21.** 30 rd. 15 ft. 2 in. **22.** 3 sq. yd. 6 sq. ft. **23.** 65 sq. rd. 222$\frac{1}{4}$ sq. ft. **24.** 174 A. 148 sq. rd.

Page 76. — 3. $\frac{27}{8}$. **4.** $\frac{60}{8}$. **5.** $\frac{93}{8}$. **6.** $\frac{104}{10}$. **7.** $\frac{262}{8}$. **8.** $\frac{144}{4}$. **9.** $\frac{197}{7}$. **10.** $\frac{592}{12}$. **11.** $\frac{408}{11}$. **12.** $\frac{647}{11}$. **13.** $\frac{501}{5}$. **14.** $\frac{283}{11}$. **15.** $\frac{565}{5}$. **16.** $\frac{716}{7}$. **17.** $\frac{1815}{14}$. **18.** $\frac{417}{11}$. **19.** $\frac{775}{13}$. **20.** $\frac{1434}{15}$. **21.** $\frac{1232}{16}$. **22.** $\frac{1077}{12}$. **23.** 6$\frac{1}{4}$. **24.** 7$\frac{3}{5}$. **25.** 9$\frac{1}{7}$. **26.** 7$\frac{1}{5}$. **27.** 7$\frac{5}{11}$. **28.** 4$\frac{1}{2}$. **29.** 7$\frac{1}{4}$. **30.** 8$\frac{7}{8}$. **31.** 11$\frac{1}{4}$. **32.** 12$\frac{3}{5}$. **33.** 9$\frac{7}{12}$. **34.** 10$\frac{7}{11}$. **35.** 10$\frac{3}{4}$. **36.** 19$\frac{4}{11}$. **37.** 27$\frac{7}{8}$. **38.** 23$\frac{1}{2}$. **39.** 24. **40.** 22$\frac{3}{5}$. **41.** 26$\frac{3}{4}$. **42.** 28$\frac{3}{4}$.

Page 77. — 1. About 145 mi. **2.** About 205 mi. **3.** 3619 ft. **4.** 1206 ft. **5.** 5490 ft. **6.** About 2$\frac{1}{4}$ hr. **7.** About 95 sq. mi. **8.** 62$\frac{1}{4}$. **9.** About 24. **10.** About 10 more in Cuba.

Page 78. — 1. 90°. **2.** 90°. **3.** 30°. **4.** No. **6.** Greater. **7.** 30°. **9.** 50°. **10.** 20°. **11.** 68°. **12.** 111°.

260

Answers

Page 79. — **1.** $7\frac{1}{5}$. **2.** $3\frac{11}{15}$. **3.** 14. **4.** 24. **5.** 58. **6.** 86. **7.** 100. **8.** 168. **9.** 3. **10.** 17. **11.** $\frac{2}{15}$. **12.** $\frac{2}{7}$. **13.** 2. **14.** 5. **15.** 30. **16.** 4. **17.** 27 ¢. **18.** \$$5\frac{3}{10}$. **19.** \$$48\frac{4}{5}$. **20.** 72 ¢. **21.** $2\frac{1}{8}$ bu. **22.** $18\frac{3}{5}$ gal. **23.** $4\frac{2}{15}$ ft. **24.** 23. **25.** 40. **26.** \$$14\frac{1}{2}$. **27.** $52\frac{1}{2}$ ¢. **28.** 20.

Page 80. — **1.** 92. **2.** $49\frac{1}{2}$. **3.** $35\frac{3}{4}$. **4.** 81 ¢. **5.** \$5.43. **6.** 13, and $12\frac{1}{4}$ sq. ft. **7.** $\frac{1}{6}$. **8.** $\frac{1}{5}$. **9.** $\frac{2}{3}$. **10.** \$1. **11.** \$1. **12.** $\frac{1}{6}$. **13.** \$10,000. **14.** $\frac{3}{8}$.

Page 81. — **2.** $21\frac{1}{4}$. **3.** $47\frac{1}{8}$. **4.** $37\frac{1}{3}$. **5.** $35\frac{1}{5}$. **6.** $50\frac{3}{5}$. **7.** $43\frac{1}{6}$. **8.** $273\frac{3}{7}$. **9.** 252. **10.** $630\frac{1}{3}$. **11.** $736\frac{1}{3}$. **12.** $16\frac{7}{8}$. **13.** \$1.83. **14.** $146\frac{1}{4}$. **15.** $68\frac{1}{5}$. **16.** $104\frac{3}{8}$ mi. **17.** \$$76\frac{9}{14}$. **18.** 1528. **19.** \$1.96.

Page 83. — **1.** 8 ¢. **2.** 64 ¢. **3.** \$1.50. **4.** \$3.00. **5.** \$1124.80. **6.** \$334,910. **7.** 15 ft. **8.** 8 ft. **9.** 6 in. **10.** Tuesday. **11.** Aug. 8. **12.** Thursday.

Page 84. — **2.** 19 yr. 3 mo. 13 da. **3.** 22 yr. 2 mo. 20 da. **4.** 21 yr. 1 mo. 24 da. **5.** 22 yr. 2 mo. 17 da. **6.** 30 yr. 3 mo. 1 da. **7.** 7 yr. 2 mo. 18 da. **8.** 9 yr. 18 da. **9.** 19 yr. 2 mo. 10 da. **10.** 25 yr. 1 mo. 14 da. **11.** 24 yr. 11 mo. 6 da. **12.** 54 yr. 9 mo. 27 da. **13.** 19 yr. 8 mo. 24 da. **14.** 33 yr. 3 mo. 22 da. **15.** 9 yr. 6 da.

Page 85. — **1.** 60 sq. ft. **2.** 30 sq. ft. **3.** 90 sq. ft. **4.** 720 sq. rd. **5.** 9 rd.; 12 rd. **6.** 240 sq. rd. **7.** 54 sq. rd. **8.** $7\frac{1}{2}$. **9.** $3\frac{3}{4}$. **10.** $33\frac{3}{4}$. **11.** $18\frac{3}{4}$.

Page 86. — **1.** About $4\frac{6}{7}$ sec. **2.** About $24\frac{1}{4}$ sec. **3.** About 3360 ft. **4.** About 2240 ft. **5.** About a mile. **6.** About 840 ft. **7.** 11. **8.** About 1 hr. $28\frac{1}{3}$ min. **9.** About $11\frac{1}{4}$ sec.

Page 87. — **1.** 76 in. **2.** 368. **3.** 720. **4.** 198. **5.** 74. **6.** 4 rd. **7.** $2\frac{1}{2}$. **8.** 2300. **9.** 40 ft. **10.** $10\frac{1}{3}$.

Page 89. — **1.** $156\frac{1}{4}$. **2.** 234. **3.** 835. **4.** $283\frac{1}{3}$. **5.** $153\frac{1}{8}$. **6.** 204. **7.** $272\frac{1}{4}$. **8.** $313\frac{1}{2}$. **9.** $346\frac{1}{4}$. **10.** 1384. **11.** $1173\frac{5}{8}$. **12.** $2570\frac{1}{3}$. **13.** $5\frac{1}{8}$. **14.** 2. **15.** 4. **16.** 3. **17.** 6. **18.** 3. **19.** $\frac{2}{15}$. **20.** 7. **21.** $12\frac{3}{14}$. **22.** 23. **23.** 17. **24.** 64. **25.** 84 sq. ft. **26.** 91 sq. ft. **27.** 125 sq. ft. **28.** 48 sq. ft. **29.** $32\frac{1}{2}$ sq. ft. **30.** 50 sq. ft. **31.** 8 ft. **32.** 9 ft. 4 in. **33.** 2 ft. 10 in. **34.** 9 ft. 2 in. **35.** 14 ft. **36.** 6 ft. 6 in.

Page 92. — **1.** 99 sq. ft. **2.** 456. **3.** $16\frac{1}{2}$ sq. yd. **4.** 1400. **5.** 600. **6.** $50\frac{2}{3}$.

Page 95. — **4.** $18\frac{1}{6}$. **5.** $32\frac{4}{5}$. **6.** $77\frac{2}{3}$. **7.** $135\frac{9}{20}$. **8.** $173\frac{1}{20}$. **9.** $4\frac{1}{6}$. **10.** $11\frac{1}{15}$. **11.** $4\frac{4}{5}$. **12.** $7\frac{7}{15}$. **13.** $15\frac{17}{18}$. **14.** $3\frac{13}{20}$. **15.** $5\frac{1}{20}$. **16.** $14\frac{11}{20}$. **17.** $21\frac{3}{5}$. **18.** $43\frac{1}{5}$. **19.** 38. **20.** $34\frac{1}{6}$. **21.** 74. **22.** 77. **23.** 270. **24.** 322. **25.** $36\frac{1}{3}$. **26.** $117\frac{1}{6}$. **27.** $187\frac{1}{15}$. **28.** $308\frac{1}{20}$. **29.** $676\frac{7}{20}$. **30.** $423\frac{4}{5}$. **31.** $3\frac{2}{5}$. **32.** $4\frac{1}{15}$. **33.** 4. **34.** 3. **35.** 11. **36.** 44. **37.** 2. **38.** 3.

Page 98. — **1.** 8. **2.** 27. **3.** 24. **4.** 8. **5.** 1728. **6.** 27. **7.** $187\frac{1}{2}$.

Answers

8. 64.   9. 4000.   10. About 1750 lb.   11. About 6 oz.   12. About 10 lb.
13. About 1203 lb.   14. About 5 oz. less.   15. About 422 lb.

Page 99. — 1. 69.   2. 113.   3. $105\frac{1}{2}$.   4. $185\frac{7}{8}$.   5. $141\frac{7}{15}$.   6. $133\frac{7}{8}$.
7. $653\frac{1}{4}$.   8. $1072\frac{7}{15}$.   9. $583\frac{9}{16}$.   10. $1251\frac{1}{4}$.   11. $889\frac{7}{15}$.   12. $1390\frac{11}{16}$.
13. $15\frac{1}{8}$.   14. $10\frac{7}{12}$.   15. $74\frac{11}{14}$.   16. $38\frac{1}{16}$.   17. $50\frac{7}{14}$.   18. $49\frac{1}{8}$.   19. $237\frac{1}{4}$.
20. $137\frac{7}{12}$.   21. $69\frac{1}{8}$.   22. $199\frac{11}{12}$.   23. $300\frac{5}{12}$.   24. $709\frac{1}{4}$.   25. 13,425.
26. 17,286.   27. 21,076.   28. 22,869.   29. 10,985.   30. 78,475.   31. $938\frac{7}{8}$.
32. $2762\frac{1}{2}$.   33. $4849\frac{7}{8}$.   34. $3571\frac{1}{4}$.   35. $1545\frac{1}{4}$.   36. $2413\frac{7}{12}$.   37. $49\frac{1}{4}$.
38. $45\frac{1}{8}$.   39. $190\frac{4}{15}$.   40. $91\frac{2}{15}$.   41. $97\frac{3}{16}$.   42. $31\frac{7}{12}$.   43. $189\frac{7}{8}$.
44. $142\frac{3}{15}$.   45. $72\frac{2}{15}$.   46. $36\frac{1}{16}$.   47. $52\frac{1}{17}$.   48. $102\frac{2}{15}$.

Page 100. — 3. 51.4.   4. 39.3.   5. 104.8.   6. 169.   7. 101.4.   8. 2.7.
9. 4.5.   10. 12.4.   11. 11.1.   12. 23.7.   13. 183.1.   14. 21.4.   15. 49.7.

Page 101. — 3. 3.2.   4. 37.1.   5. 68.6.   6. 52.2.   7. 78.3.   8. 96.6.
9. 238.   10. 664.3.   11. 547.5.   12. 1348.9.   13. 2380.9.   14. 5601.5.
15. 6.2.   16. 6.9.   17. 2.4.   18. 4.5.   19. .3.   20. .2.   21. .2.   22. .4.
23. 1.2.   24. 1.7.   25. 1.8.   26. 2.5.   27. 235.2.   28. 11.9 mi.   29. 67.6.
30. 2.7 in.

Page 102. — 3. 120.   4. 77.   5. $84\frac{1}{8}$.   6. $5273\frac{7}{16}$ lb.   7. $546\frac{7}{8}$ lb.
8. 64.   9. 118.8.

Page 103. — 1. 16 ft.   2. 200 ft.   3. 192.   4. 2048.   5. 2880.   6. 576.
7. \$256.   8. 486.   9. 100.   10. \$135.

Page 105. — 1. $1\frac{1}{8}$.   2. $1\frac{5}{18}$.   3. $1\frac{11}{14}$.   4. $2\frac{4}{5}$.   5. $1\frac{1}{3}$.   6. $1\frac{7}{8}$.   7. $\frac{4}{5}$.
8. $\frac{7}{18}$.   9. $\frac{7}{16}$.   10. $1\frac{7}{12}$.   11. $3\frac{11}{17}$.   12. $5\frac{11}{18}$.   13. $8\frac{2}{5}$.   14. 10.   15. 12.
16. $156\frac{1}{4}$.   17. 100.   18. 100.   19. $\frac{1}{13}$.   20. 2.   21. 16.   22. 4.   23. 4.
24. 12.   25. 512 cu. in.   26. 1728 cu. in.   27. 3375 cu. in.   28. $15\frac{5}{8}$ cu. ft.
29. $91\frac{1}{8}$ cu. ft.   30. $42\frac{7}{8}$ cu. ft.   31. 400 cu. in.   32. 221 cu. in.   33. 1110
cu. in.   34. 375 cu. ft.   35. 117 cu. ft.   36. 54 cu. ft.

Page 108. — 3. 2.   4. 72.54.   5. 209.65.   6. 150.2.   7. 247.2.
8. 271.2.   9. 6.86.   10. 68.6.   11. 5.25.   12. 2.88.   13. 485.12.
14. 782.31.   15. .06.   16. 1.27.   17. 1.56.   18. 4.41.   19. .12.   20. 1.2.
21. 12.   22. 92.   23. 24.02.   24. 240.2.   25. 25.03.   26. 250.3.
27. 7.92 in.   28. 213.84.   29. 66 ft.   30. 252.85.

Page 109. — 1. 60 ¢.   2. 77 ¢.   3. $6\frac{2}{3}$ da.   4. 25.   5. 12,960.   6. About
14 min.   7. $588\frac{1}{4}$.   8. $247\frac{1}{2}$.   9. 42.   10. 96.   11. 64.   12. 300.

Page 111. — 1. 8.   2. 48.   3. 36.   4. 64.   5. 12.   6. 14.   7. $\frac{1}{2}$.
8. 22.   9. 20.   10. 8.   11. 72.   12. 5.   13. 64.   14. $11\frac{2}{3}$.

Page 112. — 1. 1000.   2. 1728.   3. 3375.   4. 5832.   5. 9261.
6. 13,824.   7. 21,952.   8. 39,304.   9. 480.   10. 1080.   11. 4320.
12. 3840.   13. 7056.   14. 5184.   15. 6480.   16. 15,552.   17. $\frac{1}{8}$.   18. $\frac{1}{27}$.
19. $\frac{1}{216}$.   20. $\frac{1}{3197}$.   21. $\frac{1}{960}$.   22. $\frac{1}{4213}$.   23. $\frac{1}{125}$.   24. $\frac{1}{192}$.

Page 113. — 1. 13 bu. 2 pk. 6 qt. 1 pt.   2. 5 gal. 3 qt.   3. 28 yd.

2 ft. 6 in. **4.** 3 lb. 4 oz. **5.** 101¼. **6.** 3. **7.** $142.40. **8.** $15.75. **9.** 4 sq. rd. 3 sq. ft. **10.** 22. **11.** 288. **12.** 22,950. **13.** About $3\frac{9}{10}$. **14.** About 5⅛. **15.** 702⅓. **16.** 36.

Page 114.—**1.** 250 sq. ft. **2.** 2400. **3.** 2100. **4.** 7380. **5.** $158.67. **6.** 1050.

Page 115.—**2.** 1.4. **3.** 1.8. **4.** 25.9. **5.** 35.36. **6.** 1441.44. **7.** 3867.79. **8.** 127.92. **9.** 834.08. **10.** 419.04. **11.** 5683.23. **12.** 201,757.08. **13.** 560,332.96. **14.** 1012.69. **15.** 2072.35. **16.** 1193.28. **17.** 3556.36. **18.** 23,188. **19.** 33,340.86. **22.** 4. **23.** 2. **24.** 2. **25.** 3. **26.** 5. **27.** 6. **28.** 4. **29.** 5. **30.** 2. **31.** 6. **32.** 8. **33.** 5.

Page 116.—**7.** 84¢. **8.** 56¼¢. **9.** $3.56. **10.** 660. **11.** 14. **12.** $1.68. **13.** $3.10.

Page 117.—**5.** About 7 in. **6.** About 3½ in. **7.** About 7⅓.

Page 120.—**4.** 1¼. **5.** 4½. **6.** 3. **7.** 2⅓. **8.** 2¼. **9.** $1\frac{3}{7}$. **10.** 2⅖. **11.** $1\frac{1}{15}$. **12.** 2. **13.** $1\frac{1}{15}$. **14.** $3\frac{7}{9}$. **15.** $3\frac{1}{28}$. **16.** 18⅙. **17.** $14\frac{7}{16}$. **18.** $8\frac{7}{22}$. **19.** $11\frac{4}{11}$. **20.** $3\frac{8}{15}$. **21.** 9⅗. **22.** $9\frac{3}{16}$. **23.** 5⅗.

Page 121.—**1.** 14 in. **2.** 18. **3.** 65¼. **4.** 51,864 ft. **5.** $40{,}615\frac{5}{13}$. **6.** About 11 ft. **7.** 480,000. **8.** 33. **9.** 88.

Page 122.—**1.** 12 sq. in. **3.** 18 sq. in. **4.** 210 sq. ft. **5.** 38½ sq. in. **6.** Circumference about 44 ft.; area about 154 sq. ft. **7.** About 4 ft. **8.** About $12\frac{4}{7}$ sq. ft. **9.** 19.

Page 123.—**1.** 180°. **2.** 140°. **3.** 75°. **4.** 50°. **5.** 40°. **6.** 180°. **7.** 60°. **8.** 60°.

Page 124.—**1.** 50 hr. **2.** 30 hr. **3.** 4 da. 16 hr. **4.** 40 hr. **5.** 16⅔ hr. **6.** About 93¾ da. **7.** 240. **8.** 20 hr.

Page 125.—**1.** 4. **2.** 35. **3.** 24. **4.** 25. **5.** ¾. **6.** 6. **7.** 6. **8.** 64.

Page 128.—**3.** ⅔. **4.** ⅔. **5.** $\frac{5}{9}$. **6.** ⅗. **7.** ⅞. **8.** ⅓. **9.** $\frac{4}{9}$. **10.** ⅓. **11.** $\frac{7}{12}$. **12.** ⅝. **13.** ⅔. **14.** ¾. **15.** ¾. **16.** $\frac{17}{20}$. **17.** $\frac{5}{9}$. **18.** ½. **19.** ¾. **20.** ½. **21.** $\frac{8}{15}$. **22.** ⅜. **23.** ⅚. **24.** $\frac{21}{38}$. **25.** $\frac{7}{11}$. **26.** ⅘. **27.** ⅔. **28.** ¾. **29.** ⅚. **30.** $\frac{11}{19}$. **31.** ⅔. **32.** ½. **33.** $\frac{13}{17}$. **34.** $\frac{15}{17}$. **35.** ⅕. **36.** ¾. **37.** $\frac{8}{25}$. **38.** ⅜. **39.** $\frac{11}{16}$. **40.** $\frac{16}{19}$. **41.** ¾. **42.** $\frac{10}{31}$.

Page 129.—**1.** 65. **2.** 240. **3.** 144. **4.** 55. **5.** 606. **6.** £18 1*s.* **7.** £5 18*s.* 6*d.* **8.** £17 6*s.* 3*d.* **9.** £1 3*s.* 9*d.* **10.** 5. **11.** £8 6*s.* 8*d.* **12.** About $20.25. **13.** $36.10. **14.** £2 4*s.* 2*d.* **15.** £2 10*s.* 6*d.* **16.** About $2.30.

Page 130.—**1.** 576. **2.** $58.06. **3.** 20 ft. **4.** About $3\frac{1}{7}$ mi. **5.** About 352. **6.** 55°. **7.** 100°. **8.** 35. **13.** About $60.29.

Page 131.—**1.** $4722.90. **2.** $4338.67. **3.** $4218.94. **4.** $5479.20. **5.** $4382.42. **6.** $5875.62. **7.** $4859.73. **8.** $4852.83.

Page 132.—**1.** 16 ft. **2.** 64 ft. **3.** 256 ft. **4.** 16 ft. **5.** 144. **6.** 400 ft.

Page 134.—**6.** 40.73. **7.** 52.667. **8.** 3.186. **9.** .0052.

Answers

Page 135. — 1. $24. 2. $534.75. 5. $\frac{1}{12}$. 6. 49¢. 7. About 2$\frac{3}{4}$.
8. 220. 9. 6. 10. 45. 11. 9600. 12. 54. 13. 5 ft.. 14. 10 ft. 15. 27.

Page 136. — 1. 10$\frac{5}{8}$. 2. 9$\frac{1}{4}$. 3. 9$\frac{1}{4}$. 4. 6$\frac{9}{10}$. 5. 7. 6. 10$\frac{5}{8}$. 7. 8$\frac{14}{17}$.
8. 8$\frac{1}{4}$. 9. 13$\frac{1}{4}$. 10. 18$\frac{1}{4}$. 11. 14$\frac{3}{11}$. 12. 23. 13. 20$\frac{1}{6}$. 14. 16$\frac{5}{13}$.
15. 16$\frac{7}{13}$. 16. 15$\frac{1}{6}$. 17. 14$\frac{10}{17}$. 18. 12$\frac{1}{4}$. 19. 14$\frac{1}{4}$. 20. 18. 21. 15$\frac{1}{4}$.
22. 17$\frac{9}{23}$. 23. 17$\frac{9}{27}$. 24. 17$\frac{9}{10}$. 25. 18$\frac{4}{15}$. 26. 21$\frac{1}{4}$. 27. 39. 28. 22$\frac{1}{4}$.
29. $\frac{47}{10}$. 30. $\frac{98}{9}$. 31. $\frac{119}{10}$. 32. $\frac{91}{9}$. 33. $\frac{79}{9}$. 34. $\frac{182}{19}$. 35. $\frac{197}{20}$. 36. $\frac{119}{9}$.
37. $\frac{159}{10}$. 38. $\frac{329}{18}$. 39. $\frac{353}{9}$. 40. $\frac{509}{20}$. 41. $\frac{707}{9}$. 42. $\frac{551}{18}$. 43. $\frac{275}{12}$.
44. $\frac{471}{16}$. 45. $\frac{313}{10}$. 46. $\frac{727}{20}$. 47. $\frac{642}{17}$. 48. $\frac{924}{23}$. 49. $\frac{1220}{27}$. 50. $\frac{272}{9}$.
51. $\frac{465}{11}$. 52. $\frac{1991}{22}$. 53. $\frac{1442}{30}$. 54. $\frac{3143}{13}$. 55. $\frac{2147}{18}$. 56. $\frac{2882}{40}$. 57. 3$\frac{1}{15}$.
58. 5$\frac{1}{11}$. 59. 17$\frac{1}{8}$. 60. 3$\frac{2}{51}$. 61. 3$\frac{1}{11}$. 62. 21$\frac{1}{13}$. 63. 5$\frac{5}{14}$. 64. 37$\frac{1}{4}$.
65. 10$\frac{10}{13}$. 66. 7$\frac{2}{8}$. 67. 3$\frac{9}{16}$. 68. 27$\frac{11}{12}$. 69. 5$\frac{10}{21}$. 70. 45$\frac{4}{5}$. 71. 3$\frac{8}{104}$.
72. 60$\frac{2}{8}$. 73. $\frac{25}{112}$. 74. 45$\frac{4}{5}$. 75. 10$\frac{13}{18}$. 76. 10$\frac{10}{33}$. 77. 1$\frac{1889}{1989}$.

Page 138. — 1. 24. 2. 20. 3. 480. 4. 1584. 5. 2160. 6. $3.75.
7. $2.30. 8. $.25. 9. $3.20. 10. 28$\frac{4}{5}$. 11. 4¢. 12. $7.75. 13. $4.10.

Page 139. — 1. 180°. 2. 62°. 3. At 7 P.M. 4. 20 min. 5. 4°.
6. 57$\frac{5}{9}$°. 7. — 5.8°.

Page 140. — 1. 37. 2. 132. 3. Monday. 4. Sunday. 5. 6 mo. 15 da.
6. 8 mo. 6 da. 7. 16 yr. 3 mo. 13 da. 8. 7 yr. 5 mo. 6 da. 9. 11 yr. 2 da.
10. 144 ft. 11. 2 sec. 12. 144 ft. 13. About 2800 ft. 14. About 4360 ft.
15. 2354 ft.

Page 141. — 1. 704 yd. 2. 844 sq. rd. 3. 5$\frac{11}{16}$. 4. 244. 5. $99.
6. The former $1 less. 7. 64. 8. No difference. 9. 4 rd.

Page 142. — 1. 19$\frac{1}{4}$. 2. 22$\frac{13}{28}$. 3. 38$\frac{1}{8}$. 4. 47$\frac{7}{10}$. 5. 8$\frac{4}{5}$. 6. 13$\frac{7}{13}$.
7. 14$\frac{7}{24}$. 8. 17$\frac{21}{28}$. 9. 14$\frac{5}{10}$. 10. 23$\frac{17}{22}$. 11. 25$\frac{17}{11}$. 12. 15$\frac{1}{14}$. 13. 78$\frac{2}{5}$.
14. 69$\frac{1}{4}$. 15. 111$\frac{28}{25}$. 16. 137$\frac{13}{11}$. 17. 264$\frac{4}{5}$. 18. 452$\frac{1}{4}$. 19. 803.
20. 2054$\frac{4}{5}$. 22. 74$\frac{1}{2}$. 23. 211$\frac{3}{5}$. 24. 153$\frac{3}{4}$. 25. 288$\frac{14}{16}$. 26. 132$\frac{2}{11}$.
27. 73$\frac{1}{5}$. 28. 35$\frac{1}{4}$. 29. 99$\frac{8}{10}$.

Page 144. — 1. 20.09181. 2. 31.9747758. 3. .0000975. 4. .625.
5. 2.5. 6. 10. 7. .6. 8. 2.5. 9. 125.003. 10. 18.3462. 11. .39664.
12. .1468. 13. 2.312. 14. .0003. 15. .458. 16. .000025. 17. 3.1635.
18. .027. 19. 333.87552. 20. .0001122.

Page 145. — 1. 65°. 2. 81°. 3. 60°. 4. About 11 ft. 5. 4$\frac{5}{13}$.
6. 64 ft. 7. 4 sec. 8. 30. 9. 84,900. 10. 1306$\frac{1}{4}$ lb. 11. 24. 12. 26 yr.
10 mo. 5 da. 13. 170. 14. 68 yr. 8 mo. 27 da. 15. 81 yr. 7 mo. 29 da.

Page 146. — 3. 2,972,369. 4. 55,499,438. 5. 5202. 6. 9899.
7. 92,760,000. 8. 2,506,528. 9. 953,000. 10. 312,700,000.
11. 30,385,800. 12. 2000. 13. 10,000. 14. 86.5. 15. 24.5. 16. 37.25.

Page 147. — 1. 109.956 in. 2. 87.34 in. 3. 7.854 mi. 4. 22.
5. About 775. 6. 849.375 lb. 7. 7.86 oz. 8. 2.72. 9. 1.574 oz.
10. About 5. 11. .68+. 12. $20.67. 13. 4031.607+.

264

Answers

Page 148. —1. 24,900 mi. **2.** About 69 mi. **3.** 345. **4.** $535\frac{1}{2}$ mi. **5.** About 1037. **6.** About 26. **7.** About 52. **8.** $333\frac{1}{3}$ da. **9.** 1,507,968 mi. **10.** About 55,850.

Page 149. —1. 183.7059. **2.** 340.3846. **3.** 393.276. **4.** 240.721. **5.** 152.163. **6.** 71.2897. **7.** 612.30487. **8.** 1832.262. **9.** 4. **10.** 3.15. **11.** .2511. **12.** 22.737. **13.** 329.73. **14.** 27.795. **15.** 81.2866. **16.** .138816. **17.** 5.635. **18.** 4.18. **19.** .927. **20.** .24512. **21.** .000063. **22.** 12.25. **23.** 4. **24.** .14. **25.** 9.615. **26.** 12.12. **27.** 5.05. **28.** 8.5. **29.** 12.9. **30.** .49. **31.** 7.1. **32.** 175.1. **33.** 7.132. **34.** .46. **35.** .1462. **36.** 253. **37.** .048. **38.** .24. **39.** 1.815. **40.** .241.

Page 150. —1. 63. **2.** 56. **3.** \$1.46. **4.** \$1.30. **5.** 207 sq. ft. **6.** 1147 sq. ft. **7.** 5 rd. **8.** $\frac{1}{3}$. **9.** $\frac{3}{8}$. **10.** $87\frac{1}{2}$ cu. ft. **11.** 12,960. **12.** $289\frac{5}{12}$. **13.** 36. **14.** 330.

Page 151. —1. 28 in. **2.** 24.0352 in. **3.** 615.7536 sq. in. **4.** 168.2464 sq. in. **5.** About $\frac{7}{9}$. **6.** About $\frac{2}{9}$. **7.** 12 in. **8.** 169 sq. in. **9.** 153.9384 sq. in. **10.** About 28 ft.

Page 152. —1. 40. **2.** $68\frac{1}{4}$. **3.** $53\frac{1}{3}$. **4.** 18. **5.** 37. **6.** 20. **7.** 42. **8.** 64. **9.** 40. **10.** 40. **11.** 47.124 in. **12.** 89.5356 in. **13.** 13.36+ in. **14.** 23.73+ in. **15.** 17.2788 ft. **16.** 5.57+ ft. **17.** 4.1888 ft. **18.** 201.0624 sq. in. **19.** 1256.64 sq. in. **20.** 7.957+ ft. **21.** 49.731+ sq. ft. **22.** 1963.5 sq. ft.

Page 156. —22. $2\frac{14}{15}$. **23.** $2\frac{11}{20}$. **24.** $2\frac{29}{56}$. **25.** $6\frac{16}{35}$. **26.** $3\frac{19}{27}$. **27.** $4\frac{8}{45}$. **28.** $6\frac{13}{46}$. **29.** $3\frac{11}{14}$. **30.** $7\frac{7}{57}$. **31.** $6\frac{39}{74}$. **32.** $6\frac{24}{55}$. **33.** $42\frac{27}{40}$. **34.** $4\frac{79}{162}$. **35.** $28\frac{34}{63}$. **36.** $28\frac{4}{7}$. **37.** $16\frac{24}{61}$. **38.** $154\frac{2}{5}$. **39.** $17\frac{49}{120}$. **40.** $266\frac{2}{3}$. **41.** $1\frac{391}{450}$. **42.** $2\frac{7}{9}$. **43.** $\frac{1}{7}$. **44.** $\frac{25}{196}$. **45.** 28. **46.** $22\frac{2}{5}$. **47.** 19. **48.** $25\frac{1}{3}$. **49.** $18\frac{3}{7}$. **50.** 15. **51.** $12\frac{33}{56}$. **52.** 20. **53.** $12\frac{1}{2}$. **54.** 48. **55.** $5\frac{4}{9}$. **56.** 185. **57.** 40. **58.** 9. **59.** $37\frac{1}{2}$. **60.** $\frac{69}{400}$. **61.** $\frac{3}{28}$. **62.** $\frac{29}{300}$. **63.** $40\frac{29}{37}$.

Page 157. —1. 66 ft. **2.** 12. **3.** 44. **4.** $16\frac{1}{2}$. **5.** $14\frac{1}{4}$. **6.** 22. **7.** 252. **8.** \$11.75.

Page 158. —2. 2, 3, 7. **3.** 2, 2, 2, 3, 3. **4.** 2, 2, 2, 3, 3, 3. **5.** 2, 2, 2, 2, 3, 3, 3. **6.** 2, 2, 3, 3, 3. **7.** 2, 2, 2, 2, 3, 5. **8.** 2, 2, 3, 5, 5. **9.** 2, 2, 3, 3, 3, 3. **10.** 2, 2, 3, 3, 3, 3, 3. **11.** 2, 2, 2, 2, 2, 3, 5. **12.** 2, 2, 3, 3, 11. **13.** 2, 2, 3, 3, 13. **14.** 21. **15.** 12. **16.** 20. **17.** 48. **18.** 11. **19.** 96.

Page 159. —1. 880. **2.** $2\frac{1}{2}$. **3.** 20 min. **4.** About $5\frac{1}{25}$. **5.** 2612 ft. greater. **6.** 1020 oz. **7.** $7\frac{1}{2}$. **8.** $44\frac{13}{48}$. **9.** $15{,}937\frac{1}{2}$ lb. **10.** $12{,}115\frac{5}{32}$ lb.

Page 160. —1. 32. **2.** 448. **3.** 14. **4.** 3. **5.** 8. **6.** 16. **7.** 24. **8.** 128. **9.** 736. **10.** $\frac{1}{4}$. **11.** $2\frac{1}{2}$. **12.** $7\frac{1}{2}$. **13.** $6\frac{9}{16}$.

Answers

Page 161. — **3.** 50. **4.** 17,000. **5.** 230. **6.** 40. **7.** 2500. **8.** .05. **9.** .02125. **10.** 1.7633+. **11.** 1.3. **12.** 20,000. **13.** 34.02. **14.** .1356435. **15.** 2587.4. **16.** 40.33. **17.** 3531.075. **18.** 69.23+.

Page 162. — **1.** 25\frac{1}{3}$. **2.** 12\frac{1}{2}$. **3.** $33. **4.** $\frac{1}{4}$. **5.** $\frac{7}{8}$. **6.** $32. **7.** 10$\frac{2}{3}$. **8.** $\frac{7}{24}$. **9.** 192. **10.** 60. **11.** $40. **12.** $24. **13.** $20.

Page 164. — **1.** 15. **2.** $\frac{1}{2}$. **3.** 21. **4.** 8$\frac{1}{4}$. **5.** 21$\frac{11}{35}$. **6.** 2. **7.** $\frac{1}{2}$. **8.** 1$\frac{1}{5}$. **9.** 1$\frac{2}{5}$. **10.** 5$\frac{2}{3}$. **11.** 19. **12.** 12$\frac{1}{3}$. **13.** 25$\frac{3}{16}$. **14.** 100. **15.** 21. **16.** $\frac{1}{9}$. **17.** 9. **18.** 6. **19.** $\frac{5}{108}$. **20.** 64. **21.** $\frac{27}{4913}$. **22.** $\frac{25}{144}$. **23.** 4. **24.** 9.

Page 165. — **8.** $\frac{15}{30}$, $\frac{9}{30}$, $\frac{2}{30}$, $\frac{7}{30}$. **9.** $\frac{24}{36}$, $\frac{8}{36}$, $\frac{14}{36}$, $\frac{5}{36}$. **10.** $\frac{6}{42}$, $\frac{7}{42}$, $\frac{14}{42}$, $\frac{1}{42}$. **11.** $\frac{6}{18}$, $\frac{15}{18}$, $\frac{8}{18}$, $\frac{9}{18}$. **12.** $\frac{5}{40}$, $\frac{18}{40}$, $\frac{6}{40}$, $\frac{9}{40}$. **13.** $\frac{16}{48}$, $\frac{40}{48}$, $\frac{18}{48}$, $\frac{3}{48}$. **14.** $\frac{24}{30}$, $\frac{25}{30}$, $\frac{14}{30}$. **15.** $\frac{8}{40}$, $\frac{15}{40}$, $\frac{14}{40}$. **16.** $\frac{25}{50}$, $\frac{15}{50}$, $\frac{4}{50}$. **17.** 1$\frac{3}{8}$. **18.** 1$\frac{1}{18}$. **19.** 1$\frac{2}{5}$. **20.** 1$\frac{7}{15}$. **21.** $\frac{4}{5}$. **22.** 1$\frac{1}{3}$. **23.** 16$\frac{1}{10}$. **24.** 14$\frac{1}{18}$. **25.** 27$\frac{7}{40}$. **26.** $\frac{1}{2}$. **27.** 1$\frac{2}{25}$. **28.** 1$\frac{23}{50}$. **29.** 1$\frac{1}{24}$. **30.** 1$\frac{17}{48}$. **31.** 1$\frac{1}{12}$. **32.** 19$\frac{27}{50}$. **33.** 21$\frac{37}{48}$. **34.** 27$\frac{19}{24}$. **35.** $\frac{1}{10}$. **36.** $\frac{1}{18}$. **37.** $\frac{1}{30}$. **38.** $\frac{3}{40}$. **39.** $\frac{13}{40}$. **40.** $\frac{19}{120}$. **41.** 3$\frac{3}{10}$. **42.** 5$\frac{7}{20}$. **43.** 4$\frac{13}{60}$.

Page 166. — **16.** 74. **17.** 105. **18.** $15. **19.** $35. **20.** $15.

Page 167. — **1.** 60$\frac{1}{8}$ sq. ft. **2.** 168 sq. in. **3.** 95$\frac{5}{8}$ sq. ft. **4.** 480 sq. ft. **5.** 884. **6.** 28.2744 ft. **7.** 31.83+ ft. **8.** 5.3+ hr. **9.** 78.54 sq. yd. **10.** 113.0976 sq. ft. **11.** 30.9024 sq. ft. **12.** 250 ; 350 ; 980. **13.** 2 ; 18 ; 51. **14.** 925 ; 1634. **15.** 44 ; 40$\frac{5}{8}$.

Page 168. — **2.** 650. **3.** 345. **4.** 3465. **10.** 30¢. **11.** 10¢. **12.** $7.20. **13.** $2. **14.** 25. **15.** 300.

Page 169. — **2.** 3. **3.** 15. **4.** 16. **5.** 50. **6.** 136. **7.** 575. **8.** 992. **9.** 946. **10.** 2422. **11.** 3315. **13.** 400. **14.** 300. **15.** 1500. **16.** 1200. **17.** 800. **18.** 2000. **19.** 2000. **20.** 400. **21.** 1800. **22.** 3000. **23.** $\frac{1}{2}$. **24.** 3. **25.** 40. **26.** $\frac{1}{4}$. **27.** 32$\frac{1}{2}$. **28.** 42. **29.** 9$\frac{1}{2}$. **30.** 18. **31.** 5000. **32.** 12,000. **33.** 12,000. **34.** 8800. **35.** 2000. **36.** 8200. **37.** 22,500. **38.** 10,300.

Page 170. — **1.** 180°. **2.** 2. **3.** 67$\frac{1}{2}$°. **4.** 47$\frac{1}{2}$°. **5.** 60°. **6.** 60°. **7.** 65$\frac{7}{8}$°. **8.** 36 ; 85. **9.** 30,600 ; $10. **10.** $300. **11.** 350. **12.** 123. **13.** 336. **14.** 1200 lb. **15.** $45.

Page 171. — **1.** 2$\frac{1}{4}$. **2.** 1$\frac{8}{15}$. **3.** 1$\frac{13}{18}$. **4.** 1$\frac{2}{3}$. **5.** $\frac{5}{9}$. **6.** $\frac{3}{8}$. **7.** $\frac{3}{4}$. **8.** 1$\frac{11}{40}$. **9.** 1$\frac{15}{32}$. **10.** 1$\frac{2}{7}$. **11.** $\frac{19}{30}$. **12.** $\frac{8}{21}$. **13.** $\frac{1}{3}$. **14.** $\frac{13}{24}$. **15.** $\frac{3}{20}$. **16.** $\frac{1}{18}$. **17.** $\frac{5}{14}$. **18.** $\frac{1}{21}$. **19.** $\frac{13}{35}$. **20.** $\frac{17}{48}$. **21.** $\frac{4}{7}$. **22.** $\frac{3}{10}$. **23.** $\frac{5}{12}$. **24.** $\frac{6}{7}$. **25.** $\frac{13}{31}$. **26.** 4022$\frac{5}{8}$. **27.** 6881$\frac{3}{8}$. **28.** 26,709$\frac{1}{7}$. **29.** 5687$\frac{1}{2}$. **30.** 31,332$\frac{8}{9}$. **31.** $\frac{3}{10}$. **32.** $\frac{3}{13}$. **33.** $\frac{3}{31}$. **34.** 3. **35.** 4. **36.** 1$\frac{1}{4}$. **37.** 1$\frac{1}{4}$. **38.** 1$\frac{3}{8}$. **39.** 2$\frac{2}{5}$. **40.** 3$\frac{5}{9}$.

Page 173. — **11.** 2$\frac{1}{8}$. **12.** 1$\frac{5}{9}$. **13.** 2$\frac{19}{20}$. **14.** 1$\frac{5}{24}$. **15.** $\frac{15}{16}$. **16.** 9$\frac{4}{9}$. **17.** 9$\frac{7}{10}$. **18.** 15$\frac{1}{4}$. **19.** 17$\frac{1}{28}$. **20.** 33$\frac{13}{15}$. **21.** $\frac{1}{6}$. **22.** $\frac{4}{15}$. **23.** $\frac{5}{24}$. **24.** $\frac{7}{20}$. **25.** $\frac{1}{12}$. **26.** 9$\frac{7}{15}$. **27.** 8$\frac{13}{15}$. **28.** 8$\frac{1}{2}$. **29.** 9$\frac{7}{9}$. **30.** 6$\frac{3}{5}$. **31.** $\frac{3}{20}$. **32.** $\frac{13}{15}$.

Answers

33. $\frac{8}{40}$. 34. $15\frac{5}{16}$. 35. $19\frac{23}{36}$. 36. $1\frac{3}{8}$. 37. $1\frac{3}{8}$. 38. $4\frac{3}{4}$. 39. $1\frac{1}{15}$.
40. $1\frac{1}{2}$. 41. $6\frac{1}{4}$. 42. $3\frac{81}{104}$. 43. $\frac{11}{18}$. 44. $\frac{41}{56}$. 45. $\frac{22}{150}$. 46. $1\frac{11}{12}$. 47. $1\frac{4}{7}$.
48. $51\frac{1}{2}$. 49. $\frac{27}{520}$. 50. $\frac{155}{418}$.

Page 175.—3. $60.55. 4. $.87. 5. $7.60. 6. $19.02. 7. $3.81.
8. $20.76. 9. $.60. 10. $.03. 11. $.88. 12. $4.74. 13. $2.43.
14. $2.44.

Page 177.—1. 100. 2. 600. 3. 1620. 4. 1200. 5. 310. 6. 941.
7. 522. 8. 4983. 9. 1912. 10. 2080. 11. 3159. 12. 5785. 13. 730.
14. 1888. 15. 3136. 16. 4308. 17. 2620. 18. 1040. 19. 1600.
20. 3600. 21. 3280. 22. 15,610. 23. 6100. 24. 3220. 25. 126.
26. 729. 27. 287. 28. 639. 29. 1550. 30. 2979. 31. 1500. 32. 2550.
33. 6354. 34. 5600. 35. 7758. 36. 2955. 37. 3750. 38. 2195.
39. 2344. 40. 1150. 41. 2350. 42. 1075. 43. 30,720. 44. 16,992.
45. 17,586. 46. 19,384. 47. 7768. 48. 5136.

Page 178.—1. 98¢. 2. 75¢. 3. $7.75. 4. 75¢. 5. $7. 6. $6.
7. $77\frac{1}{12}$. 8. $29\frac{1}{11}$. 9. $4\frac{1}{2}$. 10. 10. 11. $2.

Page 179.—1. 5. 2. $50; $250. 3. $\frac{1}{8}$; $12\frac{1}{2}\%$. 4. 20%. 5. $\frac{1}{8}$.
6. 25%. 7. 85%; $42\frac{1}{2}$¢. 8. 280; 600. 9. 9; 4; 12. 10. 30; 46; 32.
11. $5.05. 12. $8.40. 13. $5\frac{1}{2}$ in. 14. 15.

Page 180.—1. 91 lb. 2. 2 bu. 5 qt. 3. 3 lb. 3 oz. 4. 2 gal. 2 qt.
5. $49. 6. $50. 7. 60. 8. 28,500 lb. 9. $5394. 10. 2544. 11. 90¢.
12. 90¢. 13. $1209.

Page 181.—1. 37.6992 ft. 2. 30. 3. 113.0976. 4. 57. 5. About
$437\frac{1}{5}$. 6. $3.62. 7. 21. 8. 684.

Page 183.—1. 27 da. 2. 15. 3. $7.50. 4. 324. 5. 8425. 6. 463.
7. 784. 8. $\frac{1}{264}$. 9. $\frac{1}{2}$. 10. $19\frac{7}{17}$. 11. About 90. 12. $58\frac{3}{4}$. 13. $8000.
14. $4500.

Page 184.—1. 18 in. long, 8 in. wide. 2. 360. 3. 228 sq. ft. 4. 259
sq. ft. 5. 486 sq. ft. 6. 84. 7. 576. 8. 12. 9. 9 ft. × 4 ft. or
18 ft. × 2 ft. 10. $99. 11. 5, 10, and 15. 12. $40 and $60. 13. $20,
$80, and $100.

Page 186.—1. 12. 2. 864. 3. $156.80. 4. $26\frac{3}{4}$. 5. 480. 6. $34\frac{1}{4}$.
7. $13\frac{1}{4}$. 8. $127.50. 9. $\frac{1}{8}$. 10. $12\frac{1}{2}\%$. 11. 307.8768 sq. ft. 12. 300
sq. ft. 13. 256.1232 sq. ft.

Page 188.—2. $\frac{7}{8}$. 3. $\frac{11}{12}$. 4. $\frac{4}{5}$. 5. $\frac{5}{8}$. 6. $\frac{5}{8}$. 7. $\frac{1}{13}$. 8. $\frac{4}{5}$. 9. $\frac{11}{13}$.
10. 119 bu. 1 pk. 6 qt. 11. 261 ft. 12. 89 yd. 2 ft. 2 in. 13. 30 mi.
289 rd. $\frac{1}{2}$ yd. 14. 452 sq. rd. $255\frac{1}{4}$ sq. ft. 15. 1143 cu. ft. 146 cu. in.
16. 1942 lb. 8 oz. 17. 9. 18. 9. 19. 7¢. 20. 25. 21. $10\frac{3}{4}$.
22. $67\frac{9}{13}$. 23. 14.304.

Page 189.—1. 450 lb. 2. About $11\frac{1}{2}$ oz. 3. About $6\frac{4}{5}$. 4. $64\frac{1}{4}$ lb.
5. About 30. 6. About 20. 7. 8.78. 8. 7.29.

Answers

Page 190.—9. $\frac{1}{4}$; 25%. 10. $\frac{1}{5}$; 20%. 11. 5. 12. 10%. 13. 10%.
14. 25%. 15. 20%. 16. 5%. 17. 5%.

Page 191.—11. 5000. 12. 10,500. 13. $1500. 14. $1200. 15. 15.
16. 10 oz. 17. $17\frac{1}{2}$. 18. 400. 19. 7500.

Page 192.—1. $2\frac{1}{4}$. 2. $1\frac{2}{3}$. 3. $2\frac{1}{4}$. 4. $1\frac{11}{15}$. 5. $1\frac{13}{18}$. 6. $1\frac{11}{16}$. 7. $1\frac{7}{10}$.
8. $1\frac{1}{13}$. 9. $1\frac{5}{13}$. 10. $1\frac{15}{18}$. 11. $\frac{13}{15}$. 12. $\frac{11}{12}$. 13. $12\frac{1}{14}$. 14. $10\frac{19}{21}$.
15. $11\frac{13}{16}$. 16. $5\frac{5}{75}$. 17. $35\frac{21}{23}$. 18. $27\frac{1}{65}$. 19. $\frac{7}{8}$. 20. $\frac{7}{14}$. 21. $\frac{11}{30}$. 22. $\frac{7}{12}$.
23. $\frac{9}{35}$. 24. $1\frac{4}{5}$. 25. $4\frac{4}{5}$. 26. $6\frac{11}{14}$. 27. $20\frac{1}{4}$. 28. $21\frac{1}{4}$. 29. $22\frac{1}{2}$. 30. $22\frac{4}{11}$.
31. $1\frac{4}{5}$. 32. $8\frac{21}{24}$. 33. $11\frac{41}{42}$. 34. $35\frac{11}{12}$. 35. $23\frac{21}{24}$. 36. $18\frac{11}{14}$. 37. 2.
38. $4\frac{4}{5}$. 39. 3. 40. 16. 41. 1316. 42. 20. 43. $\frac{1}{4}$. 44. $\frac{10}{18}$. 45. $\frac{5}{11}$.
46. $\frac{21}{44}$. 47. $1\frac{1}{4}$. 48. $\frac{6}{16}$. 49. 26. 50. $48\frac{4}{5}$. 51. $2\frac{5}{17}$. 52. $18\frac{1}{4}$.
53. $201\frac{7}{8}$. 54. $615\frac{11}{16}$. 55. $\frac{9}{13}$. 56. 48. 57. 10. 58. $3\frac{2}{3}$. 59. 14. 60. $\frac{4}{5}$.
61. $3\frac{11}{24}$. 62. $11\frac{4}{5}$. 63. $1\frac{17}{21}$. 64. $\frac{82}{189}$. 65. $35\frac{5}{11}$. 66. $6\frac{21}{23}$. 67. $\frac{4}{37}$.
68. 100. 69. $5\frac{11}{12}$. 70. $2\frac{7}{16}$. 71. $4\frac{32}{123}$. 72. $23\frac{21}{51}$.

Page 193.—1. $4.62. 2. 75 ft. 3. 2 gal. 2 qt. 1 pt. 4. 39 mi.
5. 12 hr. 6. 4 hr. 7. $\frac{1}{12}$; $72. 8. $172.80. 9. $96. 10. $7.
11. 348. A. $148\frac{1}{4}$ sq. rd.

Page 194.—1. 39. 2. Friday. 3. Tuesday. 4. 52 wk. 1 da.
5. Monday. 6. 169 or 170. 7. 51 yr. 10 mo. 6 da. 8. 9 yr. 6 mo.
24 da. 9. $25\frac{3}{4}$. 10. 27. 11. $\frac{1}{4}$. 12. 8. 13. $26\frac{3}{4}$. 14. 132 ft. 15. 464.
16. 352.

Page 195.—5. 110%. 6. 2; 200. 7. 80%. 8. 3; 300. 9. 150%; 10.
10. $4. 11. 50¢. 12. 60¢. 13. $\frac{2}{5}$; 40%.

Page 196.—1. 4. 2. 1000. 3. 5 in. 4. 8 in. 5. 16 ft. 6. 27.
7. 16 ft. 8. $\frac{3}{4}$. 9. $\frac{1}{4}$. 10. 2. 11. 245. 12. $237\frac{1}{2}$ lb. 13. 150 lb.

Page 197.—1. 923.85. 2. 578.604. 3. 119.889. 4. 29.1276.
5. 5.069. 6. 2.15988. 7. .39375. 8. .148302. 9. .50547. 10. .034596.
11. .001508. 12. .003564. 13. 3456. 14. 16,512. 15. 723.4. 16. 930.5.
17. 3379. 18. 25. 19. 6.4. 20. 61.25. 21. 307,482. 22. 120.9516.
23. 22.7766. 24. 42.546515. 25. 1.369. 26. 4.739. 27. 36.18. 28. .082.
29. 123.071. 30. 550.583. 31. 1.786. 32. 7.535. 33. 704.258. 34. 12.533.
35. 375.103. 36. 31.634. 37. .432. 38. .213. 39. 2.572. 40. 81,314.5.
41. 31,416.2. 42. 1,032,364. 43. .124. 44. 129.582. 45. .006.
46. 618,203.333. 47. .082. 48. 1734.915.

Page 199.—1. 21. 2. 10 ft. 3. $63\frac{1}{4}$ sq. ft. 4. $87\frac{1}{2}$ sq. ft. 5. 72 sq. in.
6. $72\frac{11}{12}$. 7. $402\frac{1}{4}$ sq. ft.

Page 201.—5. $2.25. 6. 70 lb. 7. $6.50. 8. $1355\frac{11}{12}$. 9. 73¢.
10. $58\frac{21}{24}$ min. 11. $5\frac{1}{2}$¢. 12. $1.23. 13. $4\frac{1}{2}$ mi. 14. $26\frac{1}{16}$ mi.

Page 202.—5. 20%. 6. 200%. 7. $12\frac{1}{2}$%. 8. $16\frac{2}{3}$%. 9. 24. 10. 2%.
11. 25 rd. 12. 32 da.; $5\frac{1}{2}$ da. 13. 25. 14. $21\frac{1}{4}$ mi. 15. $16\frac{1}{4}$ mi.

Answers

Page 204. — **1.** 15 mi. **2.** 40 mi. **3.** 1 wk. 2$\frac{11}{14}$ da. **4.** 22$\frac{2}{3}$ in. **5.** 8 ft.
6. 6.16. **7.** 6.84+. **10.** 46,305. **11.** \$15,972. **12.** 55$\frac{9}{13}$ oxygen, 6$\frac{11}{13}$
hydrogen. **13.** 2000$\frac{1}{4}$.

Page 205. — **1.** 50$\rlap{/}c$. **2.** 20%. **3.** 20%. **4.** 12%. **5.** \$4. **6.** 40%.
7. 60$\rlap{/}c$. **8.** 20%. **9.** 10. **10.** 10. **11.** 20%. **12.** 25%. **13.** 10$\rlap{/}c$. **14.** 90%.

Page 206. — **1.** 12 hr. **2.** 33 mi. **3.** 6$\frac{1}{4}$ mi. **4.** 7 hr. **5.** \$1.65.
6. \$5. **7.** 10%. **8.** 20%. **9.** 10%. **10.** 20%. **11.** 16$\frac{2}{3}$%. **12.** 30%.
13. \$4. **14.** \$5. **15.** 80$\rlap{/}c$.

Page 207. — **5.** 2916.9978. **6.** 6.0135. **7.** \$187.07. **8.** 46.83.
9. 1217.299. **10.** 566.714. **11.** 222.7036. **12.** 250.0514.

Page 208. — **1.** 4 ; 8 ; 3 ; 11. **2.** 10 ; 15 ; 1000. **3.** 16 ; 16. **4.** 7$\frac{1}{8}$;
33. **5.** 4. **6.** 8 ; 6$\frac{1}{4}$. **7.** 15. **8.** 36. **9.** 576. **10.** 9. **11.** 5$\frac{71}{165}$.
12. Nearly 148. **13.** 175.

Page 209. — **1.** 361.84+. **2.** 1710. **3.** 358.07+. **4.** About 69.
5. 3$\frac{1}{2}$ min. **6.** About 11$\frac{1}{2}$ cu. ft. **7.** 9600. **8.** About 14 hr.
9. About 11$\frac{1}{6}$.

Page 210. — **4.** 48. **5.** 432 ft. **6.** \$7.20. **7.** \$5.47. **8.** \$106.33.
9. Thursday. **10.** July 11.

Page 212. — **3.** 7.68. **4.** 46.98. **5.** 227.7. **6.** 463.06. **7.** \$242.25.
8. \$84.68. **9.** \$204.10. **10.** \$358.47. **12.** 500. **13.** 800. **14.** 900.
15. 2200. **16.** \$21. **17.** \$20. **18.** \$210. **19.** \$42. **20.** 26.42.
21. 13.21. **22.** 66.05. **23.** 3.41. **24.** \$2.75. **25.** \$11.73. **26.** \$31.94.
27. \$114.17. **28.** 1600. **29.** 1800. **30.** 1683. **31.** 10,000. **32.** 2666$\frac{2}{3}$.
33. 6400. **34.** 9000. **35.** 9000.

Page 214. — **1.** 21.9912. **2.** 38.4846 sq. in. **3.** 516.7932 sq. in.
4. 43.9824 in. **5.** 307.8768. **6.** 263.8944 in. **7.** 6. **8.** 160.2216.
9. 160.2216. **10.** 89.012.

Page 216. — **4.** 15 ft. ; 17 ft. **5.** 12. **6.** 20 ; 40 ; 60. **7.** 10 ft.
8. 20 ft. **9.** 21$\frac{7}{8}$. **10.** 4$\frac{4}{5}$. **11.** 1$\frac{3}{32}$ cu. ft. **12.** About 1$\frac{1}{4}$ cu. ft. **13.** 30.

Page 217. — **7.** $\frac{1}{18}$. **8.** $\frac{5}{18}$. **9.** $\frac{5}{8}$. **10.** $\frac{3}{32}$. **11.** $\frac{2}{11}$. **12.** $\frac{3}{4}$. **13.** $\frac{1}{14}$.
14. $\frac{1}{18}$. **15.** $\frac{1}{8}$. **16.** $\frac{1}{4}$. **17.** $\frac{1}{8}$. **18.** $\frac{1}{4}$. **19.** $\frac{39}{40}$. **20.** 1$\frac{13}{15}$. **21.** $\frac{63}{110}$.
22. 2$\frac{4}{5}$. **23.** 6$\frac{11}{15}$. **24.** 3$\frac{4}{5}$. **25.** $\frac{9}{20}$. **26.** $\frac{1}{2}$. **27.** $\frac{9}{20}$.

Page 218. — **1.** 64. **2.** 36. **3.** 100. **4.** 10 in. **5.** 5 ft.

Page 219. — **14.** \$1.00. **15.** \$1.35. **16.** \$23.80. **17.** \$10.50.
18. \$2.08. **19.** \$.72.

Page 220. — **1.** 8$\frac{1}{4}$. **2.** 6$\frac{1}{4}$. **3.** 2. **4.** 3. **5.** 20. **6.** 20. **7.** 540.
8. 72. **9.** 24. **10.** 8$\frac{1}{3}$%. **11.** 6$\frac{1}{4}$%. **12.** \$12 ; 25%. **13.** \$25. **14.** \$20.
15. \$400. **16.** 25$\rlap{/}c$. **17.** \$33.75. **18.** \$12.75.

Page 221. — **9.** $\frac{1}{8}$. **10.** $\frac{1}{4}$. **11.** $\frac{1}{12}$. **12.** $\frac{1}{15}$. **13.** 18. **14.** 60.

Page 222. — **1.** 4. **2.** 12 sq. in. **3.** 12 ft. **4.** 60 sq. ft. **5.** 8 in.
6. 48 sq. in. **7.** 3 ft. **8.** About 5.2 ft. **9.** About 15.6 sq. ft.

269

Page 223. — **1.** 164. **2.** 174. **3.** $12\frac{1}{2}\%$. **4.** 40%. **5.** $36. **6.** $180. **7.** 100. **8.** 100. **9.** The first $20, the second $15. **10.** The first 90, the second 60. **11.** 4. **12.** 3. **13.** 18¢, 30¢, and 48¢.

Page 224. — **1.** $\frac{1}{2}$. **2.** $\frac{1}{4}$. **3.** $\frac{3}{20}$. **4.** $\frac{6}{25}$. **5.** $\frac{17}{100}$. **6.** $\frac{9}{20}$. **7.** $\frac{1}{8}$. **8.** $\frac{11}{40}$. **9.** $\frac{7}{50}$. **10.** $\frac{33}{200}$. **11.** $8\frac{1}{2}$. **12.** $9\frac{7}{20}$. **13.** $12\frac{7}{50}$. **14.** $15\frac{3}{50}$. **15.** $22\frac{2}{25}$. **16.** $3\frac{1}{8}$. **17.** $5\frac{1}{40}$. **18.** $16\frac{1}{125}$. **19.** $23\frac{6}{125}$. **20.** $42\frac{1}{500}$. **21.** .5. **22.** .25. **23.** .75. **24.** .2. **25.** .6. **26.** .15. **27.** .16. **28.** .14. **29.** .95. **30.** .96. **31.** 8.5. **32.** 9.25. **33.** 12.75. **34.** 16.2. **35.** 21.3. **36.** .75. **37.** 1.475. **38.** .042. **39.** 8.317. **40.** 9.15. **41.** 22.245. **42.** 51.087. **43.** 309.477. **44.** .1992.

Page 225. — **1.** $\frac{3}{4}$. **2.** $\frac{4}{5}$. **3.** $\frac{3}{7}$. **4.** $6\frac{2}{3}$. **5.** $20\frac{7}{12}$. **6.** 125. **7.** 12 in. **8.** 113.0976 sq. in. **9.** 30.9024 sq. in. **10.** 9 in. **11.** 254.4696 sq. in. **12.** 69.5304 sq. in. **13.** 12.5664 ft. **14.** 75.3984 sq. ft. **15.** 25.1328 sq. ft. **16.** 25.1328 sq. ft. **17.** 27.9253 + sq. ft. **18.** 15 ft. **19.** 5 ft. **20.** 12 ft. **21.** 8 ft. **22.** 20 ft.

Page 227. — **1.** $103\frac{1}{2}$ in. **2.** 444. **3.** 55.2. **4.** 499.2 sq. in. **5.** $158\frac{1}{3}$ sq. ft. **6.** 10 sq. ft. **7.** $193\frac{1}{2}$ sq. ft. **8.** 54.978 sq. ft.

Page 228. — **4** $.00\frac{1}{4}$; $.02\frac{1}{2}$; $.25\frac{5}{8}$. **5.** $\frac{1}{100}$; $\frac{1}{200}$; $\frac{1}{400}$; $\frac{3}{400}$. **6.** $\frac{1}{40}$; $\frac{13}{400}$; $\frac{41}{800}$. **7.** 25 lb. **8.** $2000. **9.** $1.80. **10.** $1.25. **11.** $22.91. **12.** $1.91. **13.** $12. **14.** $194.25. **15.** $25. **16.** $40,000.

Page 229. — **1.** $5. **2.** $1.05. **3.** $8.56. **4.** $23.50; $2373.50. **5.** 4%. **6.** $200. **7.** 5%. **8.** $200. **9.** $10.19.

Page 230. — **10.** $1\frac{17}{30}$. **11.** $2\frac{13}{56}$. **12.** $\frac{37}{72}$. **15.** $\frac{7}{20}$. **14.** $\frac{5}{12}$. **15.** $83\frac{3}{10}$. **16.** 14. **17.** $5\frac{17}{35}$. **18.** $\frac{17}{24}$. **19.** $24.51. **20.** $190\frac{2}{3}$. **21.** 42. **22.** $3\frac{1}{7}$.

Page 231. — **3.** 30°. **4.** 45°. **5.** 75°. **6.** 180°. **7.** 105°. **8.** 70°. **9.** 90°. **10.** 180°. **11.** 180°. **12.** They must be equal. **13.** 60°. **14.** 90°.

Page 232. — **3.** $32.50. **4.** $1742.40. **5.** 5. **6.** 16. **7.** $10. **8.** $13.76. **9.** $430.92. **10.** 5277.888. **11.** $105.56. **12.** 3750 lb. **13.** 615.7536. **14.** 2,309,076 lb.

Page 233. — **1.** $34.88. **2.** $209.47. **3.** $3\frac{1}{2}\%$. **4.** 2%. **5.** $122.40. **6.** $16. **7.** 2%. **8.** $36. **9.** $1810. **10.** 6000. **11.** $100.

Page 234. — **2.** $17\frac{1}{2}$ sq. ft. **3.** 1260 sq. ft. **4.** 2 sq. ft. **5.** 4 sq. ft. 68 sq. in. **6.** $33\frac{1}{4}$ sq. ft. **7.** 2 sq. ft. 69 sq. in.

Page 235. — **1.** $\frac{1}{3}$. **2.** $\frac{1}{5}$. **3.** $\frac{25}{42}$. **4.** $\frac{9}{16}$. **5.** $\frac{6}{25}$. **6.** $\frac{9}{40}$. **7.** 1. **8.** $\frac{25}{112}$. **9.** $880\frac{2}{5}$. **10.** $2643\frac{1}{4}$. **11.** $2177\frac{1}{30}$. **12.** $2419\frac{3}{28}$. **13.** $5816\frac{1}{15}$. **14.** $1514\frac{5}{99}$. **15.** $3832\frac{5}{6}$. **16.** $14,113\frac{1}{2}$. **17.** $26,536\frac{15}{32}$. **18.** $44,474\frac{4}{21}$. **19.** $29,430\frac{5}{13}$. **20.** $61,446\frac{3}{50}$. **21.** $27\frac{5}{12}$. **22.** $70\frac{10}{63}$. **23.** $62\frac{7}{20}$. **24.** $98\frac{3}{77}$. **25.** $120\frac{5}{16}$. **26.** $323\frac{1}{45}$. **27.** $3289\frac{1}{21}$. **28.** $10,707\frac{1}{2}$. **29.** $4004\frac{41}{126}$. **30.** $5888\frac{34}{35}$. **31.** $16,206\frac{3}{4}$. **32.** $23,769\frac{188}{221}$.

Answers

Page 238. — 1. 440. **2.** $40\frac{3}{8}$. **3.** 40. **4.** 160. **5.** $56\frac{1}{4}$ sq. ft. **6.** $67\frac{1}{3}$ sq. yd. **7.** $16\frac{2}{3}$%. **8.** 48¢. **9.** 50¢. **10.** 20 ft. **11.** 1200 sq. ft. **12.** $35.45.

Page 239. — **2.** 13,590 oz. **3.** About 7.86 oz. **4.** About 14.7 lb. **5.** About 14.2 lb. **6.** 3.9+ oz.

Page 240. — 1. 13.2+ lb. **2.** About 12 lb. **3.** About 6 in. **4.** About 14 in. **5.** About 7.8 lb. **6.** 144. **7.** 144 ft. **8.** 144 ft. **9.** 1872 ft. **10.** About 27,518 ft.

Page 242. — 1. $3000. **2.** $3500. **3.** 64 gal. **4.** 900 bu. **5.** 30. **6.** 30. **7.** 13%. **8.** 4%. **9.** 8%. **10.** 10¢. **11.** 16¢. **12.** $140. **13.** $1.06. **14.** $4.10.

Page 243. — 1. $40. **2.** $42.50. **3.** $12. **4.** $69.04. **5.** $800. **6.** $2000. **7.** $300. **8.** $78. **9.** $4800; $6400. **10.** $13.78.

Page 244. — 1. 8 sq. ft. 104.64 sq. in. **2.** 62.832 in. **3.** 8 sq. ft. 104.64 sq. in. **4.** 13 sq. ft. 12.96 sq. in. **5.** 55.4178 sq. in. **6.** 84.474+ sq. ft. **7.** 1413.72.

Page 245. — 1. $2\frac{11}{30}$. **2.** $1\frac{19}{42}$. **3.** $73\frac{35}{36}$. **4.** $50\frac{37}{56}$. **5.** $\frac{11}{24}$. **6.** $20\frac{1}{9}$. **7.** $35\frac{3}{20}$. **8.** $13\frac{5}{22}$. **9.** 68,825. **10.** $382\frac{1}{3}$. **11.** $\frac{3}{16}$. **12.** $\frac{1}{2}$. **13.** $8\frac{1}{3}$. **14.** 11,421. **15.** $\frac{3}{7}$. **16.** $\frac{1}{20}$. **17.** 3. **18.** $2\frac{2}{5}$. **19.** $3\frac{1}{4}$. **20.** $2\frac{8}{9}$. **21.** $35\frac{19}{40}$. **22.** $22\frac{19}{24}$. **23.** $8\frac{1}{4}$. **24.** $628\frac{1}{8}$. **25.** $14\frac{1}{12}$ mi. **26.** $65\frac{1}{8}$. **27.** 50. **28.** $26\frac{6}{49}$.

Page 246. — 1. 5 sq. ft. **2.** 7 sq. ft. 98 sq. in. **3.** $4\frac{1}{2}$ sq. ft. **4.** 6 sq. ft. **5.** 47.124 in. **6.** 353.43 sq. in. **7.** 10 sq. ft. 44.406 sq. in. **8.** 6.2832 ft. **9.** 678.5856 sq. in. **10.** 6 sq. ft. 47.064 sq. in. **11.** 12.5664 sq. ft. **12.** 25.5255 sq. ft. **13.** 31.416 ft. **14.** 314.16 sq. ft. **15.** 10.185+ ft. **16.** 325.92+ sq. ft. **17.** 25.1328 ft. **18.** 615.7536 sq. ft. **19.** 3.18+ ft. **20.** 509.29+ sq. ft.

CPSIA information can be obtained
at www.ICGtesting.com
Printed in the USA
BVHW04*1008190918
527934BV00014B/680/P